Routledge Revivals

East Indians in a West Indian Town

First published in 1986, *East Indians in a West Indian Town* explores the complex geographical, sociological and anthropological dimensions of Trinidad society before and after its political independence, by employing three sets of materials – census data, questionnaires and participant-observation records. Cartographic, humanistic and statistical approaches are combined in a historical perspective to deal with the significance of race, cultural distinctions and class in San Fernando. A major concern of the book is to examine the social complexity that lies behind geographical patterns, and to compare aggregate data with group behaviour. This book will be of interest to students of geography, sociology and anthropology.

East Indians in a West Indian Town

San Fernando, Trinidad, 1930-70

Colin G. Clarke

First published in 1986
By Allen & Unwin (Publishers) Ltd

This edition first published in 2023 by Routledge
4 Park Square, Milton Park, Abingdon, Oxon, OX14 4RN
and by Routledge
605 Third Avenue, New York, NY 10017

Routledge is an imprint of the Taylor & Francis Group, an informa business

Publisher's Note
The publisher has gone to great lengths to ensure the quality of this reprint but points out that some imperfections in the original copies may be apparent.

Disclaimer
The publisher has made every effort to trace copyright holders and welcomes correspondence from those they have been unable to contact.

A Library of Congress record exists under LCCN: 58007713

ISBN: 978-1-032-49514-9 (hbk)
ISBN: 978-1-003-39422-8 (ebk)
ISBN: 978-1-032-49519-4 (pbk)

Book DOI 10.4324/9781003394228

EAST INDIANS IN A WEST INDIAN TOWN

San Fernando, Trinidad, 1930–70

Colin G. Clarke

London
ALLEN & UNWIN
Boston Sydney

**Allen & Unwin (Publishers) Ltd,
40 Museum Street, London WC1A 1LU, UK**

Allen & Unwin (Publishers) Ltd,
Park Lane, Hemel Hempstead, Herts HP2 4TE, UK

Allen & Unwin, Inc.,
8 Winchester Place, Winchester, Mass. 01890, USA

Allen & Unwin (Australia) Ltd,
8 Napier Street, North Sydney, NSW 2060, Australia

First published in 1986

British Library Cataloguing in Publication Data

Clarke, Colin G.
East Indians in a West Indian town.
1. San Fernando (Trinidad and Tobago) – Social conditions
I. Title
972.98'303 HN246.A8
ISBN 0-04-309106-7

Library of Congress Cataloging-in-Publication Data

Clarke, Colin G.
East Indians in a West Indian town.
Bibliography: p.
Includes index.

1. Social structure – Trinidad and Tobago – San Fernando.
2. Decolonization – Trinidad and Tobago – San Fernando. 3. San Fernando (Trinidad and Tobago – Race relations. 4. San Fernando (Trinidad and Tobago – Social conditions. 5. San Fernando (Trinidad and Tobago) – Social life and customs. I. Title.
HN246.S36C58 1986 305'.097298'3 86-14088
ISBN 0-04-309106-7 (alk. paper)

Set in 10 on 11 point Bembo by Bedford Typesetters Ltd, Bedford
and printed in Great Britain by Anchor Brendon Limited, Tiptree, Essex

To Gillian,
with love and gratitude

Preface and acknowledgements

This book is, in part, a companion volume to my study *Kingston, Jamaica: urban growth and social change 1692–1962* (Clarke 1975). In that account I examined the social and urban structure of a city whose history virtually spanned the British colonial period in the Caribbean. Here I have switched from a capital to a second town – San Fernando, and the scene is Trinidad, the second largest Commonwealth Caribbean territory after Jamaica, but British only since 1815. Moreover, my focus now has changed from the long-run historical evolution of a city – its economy, social structure and spatial patterns under conditions of colonial dependency, to a more precise enquiry into the impact of decolonisation and independence on a racially segmented town during the period 1930–70.

East Indians in a West Indian town opens with a discussion of theoretical issues relating to race, culture and class and of methodological problems involving geographical and anthropological approaches to urban studies (Ch. 1), before delving into the North Indian origins of Trinidad's East Indians, their historical interaction with Creole society, and the application of the theory of pluralism to Trinidad (Ch. 2). A brief résumé is then presented of San Fernando's development: as in my Kingston book, this emphasises the rôle played by race and culture in the social structure of Caribbean cities from slavery until the mid-20th century (Ch. 3). There follows (Ch. 4) a detailed account of the social and spatial structure of San Fernando between 1930 and 1960, which draws attention to the stability of the town's social order and its geographical expression in late colonial times, and highlights the éliteness of whites and the marked social (but lack of spatial) polarity of non-white Creoles and East Indians.

Using data from a questionnaire survey, the remainder of the book is devoted to a detailed examination of the social structure of San Fernando, which penetrates beyond the spatial expression of social differences to examine their racial, cultural and class bases (Chs 5, 6 & 7). Unlike many Marxist studies of the city, however, it finds its explanations less in class analysis and class struggle than in racial and cultural segmentation and conflict. For in Trinidad, imperialism created a plural society which decolonisation has activated into a system of racial voting (Ch. 8).

Four major conclusions arise out of this study of San Fernando, each of which is significant for Trinidad as a whole and more generally for social theory: spatial proximity does not necessarily make for social integration or reduce social separation in other realms of life, contrary to common belief; household structural similarities do not imply commonalities, let alone intimacy, where East Indian parental control and racial antipathy ensure endogamy; neither religious conversion nor class mobility erodes racial segmentation; politicisation consequent on independence has made the racial segments more self-conscious and polarised.

During the course of my research I have incurred many debts. I am grateful to the Research Institute for the Study of Man, New York, for the initial grant which enabled me and my wife to work in Trinidad throughout most of 1964, and to the late Vera Rubin, its director, for her support and advice. The Institute of Social and Economic Research at the University of the West Indies at Mona, Jamaica generously provided additional field-support funds. Subsequent visits to Trinidad in 1968, 1972, 1973 and 1985 were made possible by research awards from the Latin-American Centre at the University of Liverpool and from the Nuffield Foundation.

In Trinidad I have received a warm welcome from so many people – acquaintances, respondents, friends of all social backgrounds – that it is impossible to mention everyone by name. However, Jack Harewood, the late Ena Scott-Jack, Francis Seupaul and George Sammy deserve special mention for their help and encouragement. My good friends Brahmadath Maharaj and Hamsa Ramsamooj gave up innumerable hours to accompany my wife and me on our research visits, and their generosity can never be repaid.

I am grateful to colleagues at the Universities of Liverpool and Oxford for their critical but constructive comments on several sections of this text, delivered at seminars. Also, I should like to thank Alan G. Hodgkiss, former Senior Experimental Officer, and Sandra Mather of the Drawing Office, in the Geography Department, University of Liverpool, and Jayne Lewin of the Drawing Office, School of Geography, University of Oxford, for their skill in recasting my maps.

My deepest gratitude, however, goes to Michael G. Smith of Yale University and David Lowenthal of University College, London, who have inspired and encouraged my Caribbean research for more than two decades and painstakingly criticised earlier versions of this text; and to my wife, Gillian, whose perceptive and wholehearted collaboration, in Trinidad and Britain, has been crucial to my understanding of the racial, cultural and class complexity of San Fernando.

COLIN CLARKE
Oxford, *March 1986*

Contents

Preface and acknowledgements page vii
List of tables xiii
List of plates xiv

1 *Introduction* 1

Race, culture and decolonisation 1
Race and culture in Trinidad 3
Methods and problems of fieldwork 5

2 *East Indians in Trinidad: historical and theoretical perspectives* 8

Origins of Indian immigration 8
Provenance of the Indians 9
East Indian settlement 10
Indentured labour and its aftermath 12
Caste erosion 14
Family 14
Religion 16
Society and space 18
Race and politics 21
Perspectives on the East Indians 23
Theory of pluralism 25
Conclusion 29

3 *San Fernando: from slavery to independence* 32

Population growth 34
Race and status 35
Religion 37
Town plan 37
Non-residential land use 38
Population and housing 39
Economy 41
Education 42
Politics 42
Élite groups 44

4 *San Fernando: social and spatial structure in the late colonial period* *page* 48

San Fernando in 1931 48
San Fernando in 1960 58
Linkage analysis 69
Racial groups and their characteristics 74
Conclusion 76

5 *Segmentation, stratification, race and caste* 78

Segmentation 78
Stratification 80
Race 86
Caste 89
Conclusion 97

6 *Religions, cults and festivals* 98

Religious affiliation 98
Rejuvenation of institutions 102
Hindu caste, religion and the priesthood 103
Festivals 105
Spirit world 109
Syncretism 109
Conclusion 111

7 *Household, kinship and marriage* 112

Marital condition and parental status 112
Relationship to the household head 115
Types of domestic unit 117
Placement of children 118
Union status 119
Marriage among East Indians 120
Caste and varna endogamy and exogamy 124
Conclusion 127

8 *Intersegmental association and political affiliation* 129

Inter-marriage 129
Friendship and clubbing 133
Political affiliation 135
Conclusion 141

9 Conclusion: pluralism in San Fernando and beyond *page* 143
San Fernando 143
Crisis and recovery 145
Social, cultural and structural pluralism 147
Transcending the case study 150

Appendix A Small-area census data and problems in their analysis 152
Appendix B The sample survey 153
Appendix C Tables expressing the marital and mating history of household heads and their dependents and showing dependents' relationships to heads 154
Glossary 173
Bibliography 176
Index 188

List of tables

3.1 Population by race and sex in San Fernando, 1946 *page* 34
3.2 Occupations in San Fernando, 1946 43
4.1 Indices of dissimilarity for racial and religious groups in San Fernando, 1931 51
4.2 Matrix of Spearman rank correlation coefficients for selected variables in San Fernando, 1931 55
4.3 Race and religion in San Fernando, 1960 59
4.4 Indices of dissimilarity for racial and religious groups in San Fernando, 1960 63
4.5 Matrix of Spearman rank correlation coefficients for selected variables in San Fernando, 1960 71
5.1 Relationship between occupational stratification and sample 81
5.2 Relationship between colour and occupation for Creoles (sample) 82
5.3 Relationship between educational attainment and sample 83
5.4 Why are there proportionately fewer East Indians than blacks in the police (sample)? 87
5.5 Why are there proportionately fewer East Indians than blacks in the civil service (sample)? 88
5.6 The caste of samples of Hindus and Christian East Indians in San Fernando and of Hindus in Débé (respondent and spouse) 90
5.7 Caste and class in San Fernando (sample) 95
7.1 Differing types of domestic units in the San Fernando and Débé samples 117
7.2 Placement of children less than 24 years separated from both parents in the San Fernando and Débé samples 119
7.3 Caste and marriage in San Fernando (sample) 125
8.1 The household mix in the San Fernando and Débé samples 130
8.2 Political opinions in the San Fernando and Débé samples (1) 137
8.3 Political opinions in the San Fernando and Débé samples (2) 138
9.1 Indices of dissimilarity for racial and religious groups in San Fernando and surrounding suburbs, 1970 146

List of plates

1 East Indian retailers, High Street, San Fernando.
2 Hindu cremation, the Creek, San Fernando. The Gulf of Paria is in the background.
3 First cutting of a boy's hair, *Siparu Mai* celebration, Siparia. The woman is wearing an *oronhi* (veil).
4 Wedding pole in *maro* (wedding booth), Hindu wedding. Note the mortar, pestle and rum bottle.
5 The *milap* (meeting up) of the bride and groom's entourages, Brahmin wedding, San Fernando. Many of the men are wearing the *dhoti* (loincloth).
6 *Bedi* (altar) for use at the *dwar puja* (ceremony of the gateway), Hindu marriage ceremony. The brass vessel is a *lotah*.
7 Groom wearing *maur* (crown) and visor, at *dwar puja* (ceremony of the gateway), Hindu wedding.

1 *Introduction*

Race, culture and decolonisation

In most multiracial, multi-ethnic and multicultural societies that once were part of the British Empire the process of decolonisation has exacerbated intercommunal relations. Distinctive racial, ethnic and cultural categories, whose mutual awareness under British Imperialism may have ranged from indifference to hatred, have vied to fill the power vacuum created since World War II by the promise or achievement of independence. Competition has taken many forms: communal fighting leading to partition as a prelude to independence – in the case of India and Pakistan; tribal-cum-regional rebellion – for example, Biafra's abortive secession from newly independent Nigeria; electoral rivalry between contraposed groups, such as that between Greeks and Turks in Cyprus, as a forerunner to partition.

The wish to dominate and the fear of domination lie at the root of communal discord wherever it may occur. Majorities define minorities as beyond the pale and ineligible for benefits shared only among the dominant group; minorities fear discrimination by larger groups perceived as alien or hostile. Yet some of the most blatant forms of domination and discrimination occur in societies where the minority group is in control, and the majority, as in South Africa, is the victim of political and legal disabilities, economic exploitation, and social and cultural disparagement.

I have referred to multiracial, multi-ethnic and multicultural societies, but is it possible to differentiate among racial, ethnic and cultural categories, and how do they differ from social classes? Racial, ethnic and cultural categories often overlap and are frequently confused with one another. A racial category is distinguished by virtue of perceived physical characteristics, from which moral, intellectual and other attributes and abilities are often believed to derive. Racism is the dogma that one group is condemned by nature to congenital inferiority, whereas another is favoured with superiority. A cultural category, which may or may not appear physically distinct, displays some unique cultural characteristics, for example in family or religion. Ethnic groups have a sense of identity and cohesion based on a shared tradition, including a common descent (race), language and, sometimes, dress.

In instances where racial and cultural categories are ordered in a hierarchy, the social structure may acquire many characteristics in common with class differentiation. However, classes are groups whose positions in the social stratification – usually expressed by occupational rankings – are derived from past and present divisions of labour. Marxists stress the primacy of class conflict inherent in the social relations generated by the capitalist mode of production (Harvey 1971). Followers of Weber, on the other hand, conceive of class more broadly and flexibly: to the distinction of wealth (occupation and purchasing capacity) are added the determinants prestige and power, but in reality these three facets are often correlated. Moreover, Weberians emphasise that, in recent history, class contradictions have more commonly been resolved by negotiation than by revolution (Gerth & Mills 1970).

Racial and cultural categories may be of parallel status and internally differentiated by class distinctions; equally, class consciousness may, under certain circumstances, transcend cultural and racial boundaries. Race and cultural relations form discrete sets of phenomena, but only to a limited degree; they frequently share the characteristics of other stratification systems and may coincide with other criteria of social inequality. The major distinction between race and class is that class is often mutable, whereas race, in van den Berghe's words, 'is an extreme case of status ascription making for rigid group membership' (van den Berghe 1967, 24). Yet even racial groups may change their status through conquest and colonialism or as a result of the demographic and democratic changes that have come about with the ending of imperialism.

As the British decolonised, the privileged position of whites in former colonial territories was undermined and expatriate civil servants returned home. Previously subordinate races and cultures clashed, but less because of spontaneous chemical reaction than through competition for political power and for access to mutually desired economic resources. Unlike socio-economic class conflicts, these rivalries rarely invoked left- or right-wing ideologies. Political parties, especially those created by socially parallel groups of equal size, often selected highly pragmatic objectives and neglected liberty and equality for everyone in favour of preferential and exclusive treatment for their own followers.

In the Commonwealth Caribbean, political structures have been transformed since World War II by the introduction of universal adult suffrage. White élites, previously entrenched behind restrictive property franchises, have seen their electoral advantage disappear, and only a few individual whites have been successful at the polls, usually because of their backing by recently created, anticolonial parties. Everywhere, *coloured** political élites have emerged to fill the legislative and bureaucratic vacuum created by the departure or demise of *Creole* and expatriate whites.

Caribbean electoral politics in the 1960s were dominated by the *brown*, educated, articulate middle stratum, who saw themselves as the heirs of the ruling class (James, no date). The *black* masses were bystanders, called upon every half decade to validate the democratic system by voting at a general election. Yet despite political demotion and emigration, the social, and above all, economic power of the whites remained intact. Brown politicians, steeped in British values, proved unable to remodel the social order or restructure the economy – even where they wished to do so. However, in Trinidad as in Guyana the social consequences of colour distinctions, which are so typical of Caribbean Creole societies, have been reduced in the face of non-Creole opposition, and black (*Negro*) and coloured (*mulatto*) politicians have had to compete for electoral support with parties based on culturally distinct social segments of Indian origin.

*See the Glossary on pp. 173–5.

Race and culture in Trinidad

Trinidad and Tobago – a two-island unitary state – is located in the south-east Caribbean, off the east coast of Venezuela. Colonised by Spain from 1498 to 1797, Trinidad was then a British possession until it became independent in 1962. For more than two decades the image of Trinidadian society projected by the government has been one of racial and religious harmony. 'All o' we is one', runs a local saying, and the National Anthem assures the listener, 'Here ev'ry creed and race find an equal place'. However, the emphasis on equality and harmony fails to conceal the island's social divisions. Differences of race, colour, class and culture have a long and complex history in Trinidad, and each defines important elements in the social structure.

The principal feature of Trinidad's society is the dichotomy between Creole and East Indian. The term *Creole* has a special meaning in Trinidad. It excludes the East Indian population, together with the small Carib, Syrian, Portuguese and Chinese minorities. Creoles may be white, brown or black, and phenotype correlates with cultural and socio-economic stratification. Stratification of the Creole segment was effected during slavery, under the influence of white sugar planters and administrators. Slave emancipation in 1834–8 removed the legal distinctions between strata but left the old social hierarchy essentially intact. Within less than a decade, Indian indentured labourers were brought into the island to work on the sugar estates. Their descendants, known as East Indians, now account for just over two-fifths of the island's population. However, this element is also divided and comprises Hindus, Moslems and Christians (Harewood 1975).

The Trinidadian writer, V. S. Naipaul attributes the island's postwar expressions of racial feeling to the first elections held under universal adult suffrage in 1946. An appeal to race was a relatively simple, though inflammatory, means of drumming up electoral support. 'Then', in Naipaul's words, 'the bush lawyers and the village headmen came into their own, not only in the Indian areas but throughout the island. Then the loudspeaker van reminded people that they were of Aryan blood. Then . . . the politician, soon to be rewarded by great wealth, bared his pale chest and shouted "I is a nigger too!" ' (Naipaul 1962, 83).

Division of the population into homogeneous racial and religious regions provided scope for vote-selling on a community basis. In his hilarious but trenchant account of an imaginary, multiracial community in rural Trinidad, Naipaul explores how, in one character's words, 'election bringing out all sort of prejudice to the surface' (Naipaul 1958, 136). The East Indian candidate calculates: 'it have eight thousand votes in Naparoni. Four thousand Hindu, two thousand negro, one thousand Spanish, and one thousand Muslim. I ain't getting the Negro vote and I ain't getting a thousand Hindu votes. That should leave me with five thousand' (Naipaul 1958, 52).

Forty years later the majority of Trinidad's Creoles and East Indians still support rival political parties. All six general elections since 1956 have been won by the Creole-dominated People's National Movement, created and led until his death in 1981 by Dr Eric Williams. Trinidad's national culture is essentially Creole culture, and Creoles tend to regard East Indian political activity as 'refractory if not treasonable' (Lowenthal 1972, 175). Trinidadian elections impinge on all aspects of life. The island is small (4828 km^2) and has just over a million inhabitants; moreover, sharp political lines drawn around the racial and ethnic groups make it extremely difficult for Trinidadians to avoid being stereotyped as supporters of one or the other party.

Detailed accounts of black communities date from the classic survey of Toco by Herskovits and Herskovits (1947). More recently, studies of rural East Indians have been published by Niehoff and Niehoff (1960) and by Klass (1961). The last two monographs deal with racially homogeneous, somewhat isolated, agricultural communities situated in the west coast sugar belt, and detail a high degree of Indian cultural conservation. My intention was to examine East Indians in a totally different context – an urban area where they were numerous but lived in close contact with Creoles. The choice of study area was simple: San Fernando is Trinidad's only town of note in which East Indians account for as much as a quarter of the population. In addition to being the market centre for the southern section of the sugar belt, it is a major oil-refining centre, ranks second in the island's urban hierarchy behind Port of Spain, the capital, and already in the mid-1960s provided an indication of what the rest of Trinidad was to be like 20 years later.

In 1960, on the eve of Trinidad's independence, San Fernando had a population of almost 40 000. Creoles accounted for 70% of the inhabitants and East Indians for just over a quarter. The Creole population comprised whites (3% of the town's inhabitants), blacks (47%) and a mixed category (21%). The mixed or coloured group consisted predominantly of mulattoes, and to a lesser extent of the mixed offspring of blacks or mulattoes with Chinese, Syrians, Portuguese or Caribs. In addition, this census category included a small number of people of part-Indian origin known locally as *Douglas*. Creoles dominated San Fernando numerically; they were also proportionately more numerous in the town than in the island as a whole. Most of the Christian East Indians in San Fernando were Presbyterian. They comprised 8.5% of the town's population and outnumbered Hindus (7.6%) and Moslems (5.8%), though the converse was true for Trinidad as a whole.

This book, then, is about East Indians in San Fernando, and focuses on their social status, cultural characteristics and relations with the larger Creole population. My point of departure is that the minority status of East Indians in San Fernando, compared to their preponderence in many rural districts, may have accelerated the pace of cultural change among them and intensified their social interaction with Creoles. However, San Fernando must not be treated as an encapsulated community. Relations between East Indians in town and country deserve attention, and so, too, do links between Trinidad and India. Furthermore, by extrapolation, this study may suggest something of the degree of social integration achieved in the larger Trinidadian society, and provide perspectives on inter-group relations that will illuminate racial and ethnic circumstances of international significance.

Specifically, answers are sought to the following questions. What is the history of the social and spatial structure of San Fernando, and how do East Indians fit into it? Has education stimulated changes within and between groups to yield new forms of occupational differentiation and class alliances? How does occupational mobility differ between Creoles and East Indians? Is East Indian cultural change to be measured entirely by loss of ethnic traits or is it necessary to assess what has been learned from Creoles – white, brown and black? How far do urban Hindus and Moslems retain their familial and religious traditions? Has social change occurred through creolisation or from the acceptance of West European and North American values? Do East Indians stress their racial identity or view themselves as Hindus, Moslems or Christians? Are Christian East Indians mediators? Have urban Creoles and East Indians followed the national pattern and supported different political parties, and, if so, what impact has political competition had on social relations, especially during the decades leading up to and following independence? Finally, what light do these circumstances in San Fernando shed on Trinidadian national life?

Trinidad is not, of course, unique in its multi-ethnicity. Conditions of racial and cultural complexity are common throughout the ex-colonial Third World, and emigration from these territories to Western European and North American metropolises, since World War II, has created non-white enclaves within urban populations previously often all-white. Nevertheless, Trinidad does have direct analogues elsewhere in the Third World, where colonial plantation economies were created or sustained by heavy Indian indentured immigration during the 19th and early 20th centuries. Surinam (Gastmann 1965, Speckmann 1965, Lamur 1973) and Guyana (Despres 1967, Glasgow 1970, Greene 1974, Rodney 1981) in the Caribbean, Mauritius in the Indian Ocean (Benedict 1961, 1965, Simmons 1982) and Fiji in the Pacific (Mayer 1961, 1963, Gillion 1962, Mamak 1978) are siblings of Trinidad (Schwartz 1967a, Jayawardena 1968). Indeed, it was often a matter of chance in which colony the indentured labourers ended up (Tinker 1974). In each case, immigration produced large Indian populations that remain racially and culturally distinct from the remainder of the society, are still associated with sugar cultivation, and, except in Mauritius, generate their own political organisations and leadership. Thus, this study seeks to shed light not only on multiracial and multicultural societies, generally, but more specifically on East Indian segmentation throughout the recently independent Third World.

Methods and problems of fieldwork

My holistic approach to research in San Fernando created two fieldwork problems. How could I gather data, much of it of a behavioural kind, from a substantial urban population when the technique of participant-observation would unavoidably fail to tap more than a small fraction of the inhabitants? How was I to relate to the multiplicity of social categories in view of the segmentation between Creoles and East Indians and even Moslem–Hindu distancing?

I tried to confront the first problem by using census data and survey materials. I assembled and analysed census enumeration district data for 1931, 1960 and 1970 – a set of official, small-area, longitudinal material that is almost uniquely comprehensive for Third World towns over that timespan. These data relate to race, religion, demographic characteristics, occupational and educational achievements and, on occasion, to housing provision; they enable cartographic and statistical analyses to bring each racial and cultural category into focus and then set it in the context of the whole.

Census sources provide a temporal framework for this book which concentrates on the period 1930–70; fortunately, these data were compiled at dates that correspond closely with important events in Trinidad's recent social history. The 1931 census coincided with the onset of the depression in Trinidad, which culminated in the labour disturbance of 1937 and demands for self-government; the 1946 census was taken in the same year that adult suffrage was introduced; the 1960 census, the last colonial enumeration, was carried out 2 years before independence; the 1970 census was taken immediately after the state of emergency of that year. Furthermore, an in-depth picture of San Fernando's society at the very beginning of independence is given by my 1964 sample survey of the major racial and cultural categories in the population. A carefully tested questionnaire was applied to random samples of 890 adults selected from the Creole (211), Hindu (149), Moslem (126), Christian East Indian (256) and *Dougla* (39) populations in San Fernando and from the neighbouring, largely Hindu, village of Débé (109), located on the Oropuche Lagoon. Through this survey it is possible to make careful and penetrating comparisons between the racial and religious categories that have been sampled, and to transcend the limited material provided by the census.

However, how was I to deal with my second problem and relate to such a polyglot scene? On our arrival in San Fernando my wife and I were fortunate to rent a wooden house in a Creole neighbourhood on the western edge of the town centre. Friends and neighbours soon gave us an entrée into the Creole community, where, as expatriate whites, we were readily accepted. So fragmented is social life in San Fernando that we decided to contact East Indians via their religious bodies. The Hindus, in particular, were warm in their welcome, and we quickly built up a network of informants who brought us regular invitations to attend weddings, prayer readings, cremations, house blessings and political meetings.

Creoles and East Indians, as well as subcategories such as Moslems and lower-class blacks, usually move in separate, non-overlapping circuits, and this made feasible separate involvement with each group in turn. Neither Creoles nor East Indians really understood what we found tolerable, let alone interesting, about the other, but we were forgiven our inquisitiveness because we were patent outsiders. When Creole and East Indian friends of ours met by chance at our home they were polite to one another – but distant.

Through our multiplicity of contacts built up over a nine-month period, we constructed a detailed picture of social life in San Fernando and the East Indian villages in the surrounding sugar belt. Scenes and conversations were logged daily in our diary, and out of our growing familiarity with local behaviour and contemporary issues we compiled the questionnaire. It would be wrong to give the impression that we related equally well to all races, cultures and classes. The informants with whom we had the closest contact and whom we got to know most thoroughly were Hindu and middle-class Creole. Fortunately, the census and survey data correct any tendency to focus too exclusively on these populations.

The book, therefore, employs three sets of materials – census data, questionnaires, and participant-observation records from our diary. These sources reflect the traditional methods of geography, sociology and anthropology, and I have tried to combine them to blend cartographic, statistical and humanistic approaches within an historical perspective.

In summary, the text deals with the significance of race, cultural distinctions and class in San Fernando during the years preceding and, above all, immediately following Trinidad's independence. It explores these features spatially and socially; indeed, a major concern of the book is to examine the social complexity that lies behind geographical patterns, and to compare aggregate data with group behaviour. Analysis of the spatial data provides an aggregate chronological sequence which is subsequently amplified through the survey materials and our diary.

2 *East Indians in Trinidad: historical and theoretical perspectives*

In 1815, when Britain annexed Trinidad, the slave-based sugar economy of the West Indies had already passed its zenith. The West Indian planter class, in part absentee, was encumbered with debts which the poorly managed, worked-out estates could not pay off; and pressure for slave emancipation was mounting in the West Indies and Britain. Trinidad was not only newly acquired, but underdeveloped and underpopulated. By 1834, the year of emancipation, the Trinidad slave population stood at only 22 500 – barely one-sixth of the number in Jamaica, the largest British West Indian colony (Wood 1968).

Trinidad's emancipated blacks, like their counterparts elsewhere in the Caribbean, left the estates and penetrated the vast tracts of Crown land or settled in the towns; Port of Spain, the capital, already had more than a quarter of the island's population in 1834 (Goodenough 1976). Trinidad's planters argued that a constant and plentiful supply of cheap labour was essential to maintain the profitability of the plantation system and to enable it to expand into the substantial tracts of fertile territory that flanked the Gulf of Paria. Immigration alone could supply a controlled and growing labour force of the kind sought by the planters.

Origins of Indian immigration

In the early 1840s attempts were made to attract free American blacks and West African slaves liberated by the Royal Navy, but only a few thousand arrived in Trinidad. Small groups of Madeirans and a larger population of Chinese were introduced, but their impact upon the estate labour force was negligible. In 1844, however, the British government gave permission to Trinidad, as well as to British Guiana and Jamaica, to receive Indian indentured immigrants from the ports of Calcutta and Madras. In the first instance the cost was met from funds already standing to the credit of the colony in England. Later shipments were covered by Trinidad's general revenue and by an export duty on plantation crops plus a stamp tax paid by the planters on every indenture registered in the colony (Laurence 1971a,b, Brereton 1974a).

The first immigrant ship, the Fatel Rozack, arriving from Calcutta on 3 May 1845, landed 197 men and 28 women; six people died during the voyage. In the following three years more than 5000 Indians were distributed among sugar estates concentrated on the west coast of the island. After a brief hiatus in indentured immigration, 173 Indians arrived from Calcutta in 1853, and thereafter additional recruitments – usually numbering 1500–3000 each year –

were made until 1917, when the Indian government terminated the traffic (Roberts & Byrne 1966). During the 70 years of indentured immigration, 143 939 Indians came to Trinidad, of whom only 33 294 eventually went back to India.

During the three decades before the sugar depression and cocoa boom of the 1880s (Harrison 1979), Indian labour placed the plantations on a sound footing and enabled sugar to develop as the mainstay of the export economy (Johnson 1971). By 1866 sugar output had reached 40 000 tonnes, three times the level achieved in the late 1830s, and only once again – in 1875 – did it drop below that figure. Racial and cultural complexity was the price paid for these economic gains. The Indian element in Trinidad's population increased from 27 400 in 1871 to 70 200 in 1891, and its proportion expanded from 22 to 32% during this 20-year period (Brereton & Dookeran 1982).

Provenance of the Indians

Present-day descendants of the indentured workers are largely ignorant of the areas of India from which their forbears were recruited. A few old men and women can still recall their youth near the foothills of the Himalayas; children of indentured Indians will sometimes mention that a parent came from a specific locality such as Karsiji (Holy Benares); and a few families have maintained links with kin in India, usually by letter, very occasionally through a once-in-a-lifetime visit to the ancestral village (Naipaul 1964, 266–77). Yet the community at large lacks any awareness that the majority of indentured labourers were drawn from a quite restricted area of imperial India.

Most of Trinidad's indentured immigrants came from the United Provinces and Oudh, Bihar and Bengal, and passed through the Calcutta depot (Weller 1968, Wood 1968). Outstanding source areas were the 725-km section of the Ganges Plain between Delhi and Benares, and the area to the north of the Ganges lying between Benares and the Himalayas. Towns of the two regions such as Aligar, Mathura, Agra, Shahabad, Lucknow, Kanpur, Gonda, Fyzabad, Alahabad, Basti, Gorakhpur, Azamghar, Jaunpur, Ghazipur, Benares, Gaya and Patna constantly recur in the records as places of origin of indentured labourers. One, Fyzabad, is a place name in Trinidad; others are common family or street names.

The mere fact of crossing the sea, the *kala pani* (black water), was considered defiling for Hindus, but demographic and economic circumstances in what is now Uttar Pradesh encouraged emigration. Population densities exceeded 200 persons per hectare in many localities (Crooke 1897, 7). Sugar cane was cultivated throughout the region, but wheat was the staple of life in the west of the plain and rice in the wetter east; however, wheat farmers suffered from droughts, and rice producers were impoverished by floods. Economic changes following the introduction of British rule dislocated the traditional way of life. British manufactured goods undermined local craft industries, severely depressing hand spinning and weaving. Land fragmentation, landlessness and indebtedness throughout Uttar Pradesh had caused large numbers of high-caste Brahmins and Kshatriyas to lose property to absentee speculators by 1865 (Crooke 1897, 487; Cumpston 1953, 7).

Pestilence in 1856, famine in 1860 and cholera in 1861 helped to stimulate both cityward migration and emigration. Moreover, many fled abroad to escape the hardships and reprisals following the 1857 Indian Mutiny. Recruiting agents visited the large cities and pilgrimage centres, and enticed potential emigrants with wages often five times higher than those then current in North India. Most indentured labourers later claimed they were tricked into going abroad, but a few Indians were undoubtedly glad to leave – husbands escaping wives; women abandoned by men; widows; prostitutes; people anxious to avoid the consequences of crime; those guilty of infringing caste regulations. Furthermore, not an inconsiderable number of repatriated indentured labourers signed on for further periods in Trinidad, because it was so difficult to settle back in their villages of origin in India (Wood 1968).

Fewer than 15% of Trinidad's immigrants were Moslems. Among the Hindu majority a wide range of castes is represented in the records. Agricultural castes were greatly in demand for estate work, and together with the low castes and outcastes formed over two-thirds of the immigrants. A large number of Brahmins and Kshatriyas – many of them cultivators (Nevill 1909, 95–7) – also immigrated, and together accounted for more than 10% of Hindus. Furthermore, some high-caste Hindus may have passed themselves off as members of labouring castes to improve their acceptability to recruiters at the Calcutta depot (Wood 1968, 148). Madrassis were a small minority: only 5000 Indians embarked from Madras between 1845 and 1892, though recruitment from the south increased again in the early 1900s.

East Indian settlement

At first white Creole society viewed the Indians as transients who would leave the island once their contracts expired. In 1853, however, planter-legislators decided that, although indenture should remain a five-year term, Indians would have to work on the island a further five years before they were eligible for the free return passage; after 1895, immigrants had to contribute to the cost of their repatriation. Some Indians remained in estate employment for the period of their 'industrial residence', but others cultivated plots or turned to the crafts and trades they had once practised in India.

Natural increase ensured the permanence of the Indian community. By 1871, more than one in seven of the Indians had been born in Trinidad, and in the early 1900s the local-born outnumbered Indian immigrants: for them India would never be home. During the second half of the 19th century, racially homogeneous enclaves arose in the sugar belt south of Port of Spain and in the Naparimas around San Fernando (Fig. 2.1), and were consolidated by the high rate of Indian reproduction which outstripped that of all other groups after 1920. In Creole eyes the Indians changed from sojourners to settlers, from 'coolies' into 'East Indians' (Wood 1968, 131).

A key factor in the development of the East Indian settlement pattern was the Crown Colony's decision to release land – denied to ex-slaves in the 1840s (Blouet 1976) – for small-scale farming. In 1869, 25 time-expired Indian labourers petitioned the Governor for grants of Crown land in lieu of their

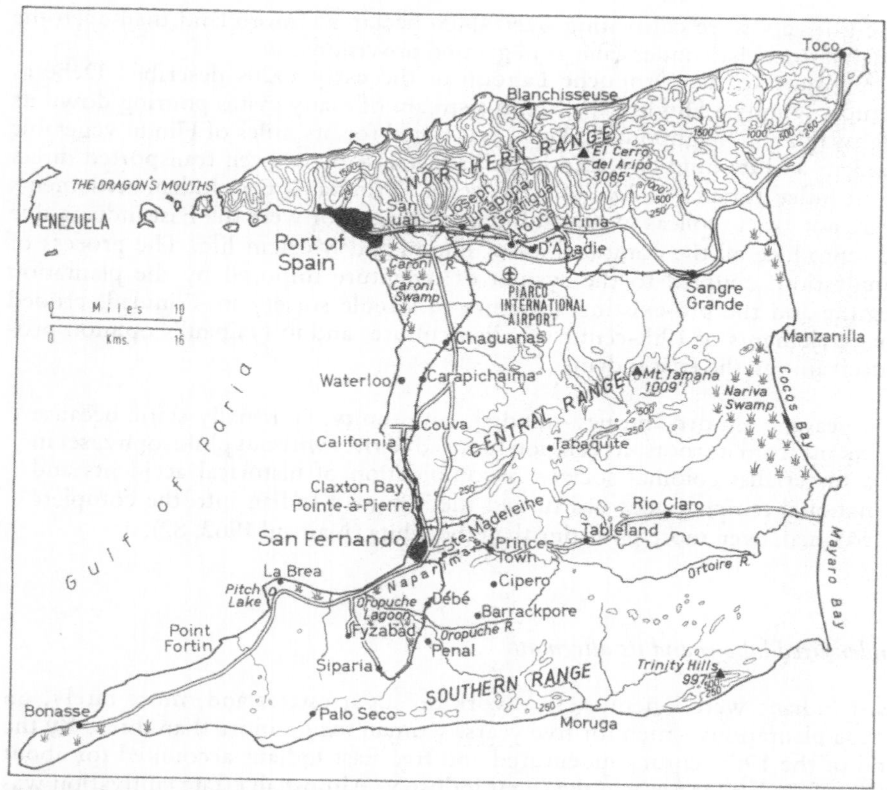

Figure 2.1 Trinidad: selected place names and communications in the late 1960s.

return passage to India. Each petitioner was granted 4 hectares at Couva and Pointe-à-Pierre. By the end of 1870, 180 Indian immigrants had commuted their return passages for land, and another 96 had bought nearly 400 hectares between them. So successful were East Indians in engrossing property that by 1916 they had 39 000 hectares in cultivation – two-thirds of it under cocoa, by then the island's main crop – and East Indian cane farmers outnumbered their Negro counterparts by four to three (Johnson 1972; Ramesar 1976b).

East Indian settlement beyond the perimeter of the plantations was originally dispersed, but after 1870 government-planned villages were built for them at Chaguanas and south of San Fernando. The planters raised little objection, and indeed often encouraged East Indian smallholding. Garden plots set out in acreages too small for subsistence were sited adjacent to sugar land, thus encouraging experienced agriculturalists to combine independent farming with plantation labour. A major feature of the East Indian peasant economy was the growing of wet rice. During the 1870s irrigated *padi* fields became prominent in the cultural landscape of the Caroni swamp and around the edge of the Oropuche Lagoon (Ramesar 1976a). By the middle of World War I, East Indian rice growers were cultivating over 4,400 hectares – more land than even the small farmers had under cane or in ground provisions.

A visitor to the Oropuche Lagoon in the early 1920s described Débé as 'almost wholly a Hindu town, with a stream of many castes pouring down its highway', and neighbouring Penal was noted for 'its miles of Hindu vegetable gardens and its mud and reed huts that seem to have been transported direct from India' (Franck 1923, 398). However, Trinidad's East Indian settlements were not exact replicas of North Indian villages; nor were the East Indians able to reproduce all the complexities of Hindu and Moslem life. The process of emigration, coupled to the system of indenture imposed by the plantation régime and the pre-existing structure of Creole society in Trinidad, eroded many elements of 19th-century Indian culture, and in Naipaul's opinion produced among the East Indians

> a peasant-minded, money-minded community, spiritually static because cut off from its roots, its religion reduced to rites without philosophy, set in a materialist colonial society: a combination of historical accidents and national temperament has turned the Trinidad Indian into the complete colonial, even more philistine than the white (Naipaul 1962, 82).

Indentured labour and its aftermath

East Indians were indentured to work on sugar estates and, more rarely, on cocoa plantations – men for five years, women for no more than three. By the end of the 19th century indentured and free East Indians accounted for about 90% of the labour force in the sugar industry. Almost all estate cultivation was done by task, and East Indians continually complained about the onerous nature of their jobs, especially digging in heavy clay and 'heading' cut cane to the delivery carts.

Estate wages for indentured labourers and freemen were equalised in 1872, but rates were invariably pitched no higher than the legal minimum and during the 1880s and 1890s they fell below it as tasks were increased. By the early 20th century, women earned between half and two-thirds the male wage: the months of the sugar crop, from January to June, remained the most remunerative period, but income plummeted during the 'dead season'.

East Indian estate workers, indentured and free, were housed in barracks or wooden ranges, each subdivided to provide miniscule cubicles for two or three single persons or a couple with children. 'The partitions between the rooms were low', Weller (1968, 59) notes, 'and by standing on a box, the occupant of one room could look or climb over the partition into the adjoining room. All the noises, conversation and odours passed easily from one room to the other'. Sanitary facilities were minimal or non-existent: indentured Indians were encouraged to defecate in the fields and under the bushes surrounding the barracks. Hookworm disease (ankylostomiasis), contracted by walking barefoot through an infected area, was endemic until estate latrines were erected in the first years of this century (Singh 1974, 58).

Health care among indentured workers was regulated by law, but conditions varied greatly in practice from estate to estate. Immigrants received regular checks by qualified medical practitioners, and all estates that employed indentured labourers had to have a hospital. Nevertheless, mortality was common among first- and second-year arrivals from India. In an attempt to improve the health of the labouring Indians, from 1879 until the end of the century, new immigrants were 'seasoned' for a year or two, during which food rations had to be supplied by their employers. Mortality rates for indentured workers declined from 37 per 1000 in 1875 to 13.5 per 1000 in 1915, but morbidity remained a problem until well beyond the indenture period, and skin ulcers, venereal disease, leprosy, malaria, yellow fever and hookworm sapped the labour force. Even the town hospitals were packed with sick, indigent Indians.

Estates in the western sugar belt became isolated residential communities of East Indians. Many plantations had their own shops, which sold groceries and provisions on credit and exacted unwarranted profits (Weller 1968, 42). More pernicious still, these stores dispensed rum 24 hours a day, and habitual heavy drinking – an escape from the trauma of emigration and indenture – became an established feature of East Indian rural life (Yawney 1969), as did the smoking of *ganja* (cannabis) (Weller 1968, 95).

Despite the legal code governing indentureship in Trinidad, social and economic life on the estates truly amounted to 'a new system of slavery' (Tinker 1974). East Indian indentured workers had only the haziest notion of their rights. To make matters worse, contractual offences of labourers against masters were deemed criminal acts, and magistrates tended to sympathise with the planters' point of view. One of the most onerous aspects of indentureship was the pass law. Not only were tickets of leave needed by indentured labourers off their estates during working hours, but time-expired East Indians also had to carry certificates of industrial residence.

Like black slaves before them, East Indian indentured workers resisted planter authority. They malingered by feigning illnesses that would admit them to the comparative comfort of the estate hospital; they absented themselves from work for days on end – though recaptured labourers were legally required to make up lost time; they burnt the sugar cane and carried out lightning strikes at the beginning of the crop season; some escaped from Trinidad altogether and settled on the uninhabited coastlands of Venezuela, which face the sugar belt on the opposite side of the Gulf of Paria (Haraksingh 1979).

The ending of indenture did not rupture the East Indians' contact with the land. A few became shopkeepers, money lenders, large landowners, or even, when the oil industry began during World War I, bonanza millionaires; they educated their children for the professions – school teaching, medicine, the law, though in the police force and the civil service (Creole strongholds) East Indians were notoriously under-represented (Smith 1974, 322): yet as late as 1946, the vast majority of East Indians remained essentially impoverished and illiterate, the backbone of Trinidad's plantation labour force, locked into a colonial employment structure that inevitably bore little or no relationship to North Indian economic circumstances or caste prescriptions (Tyson 1939; Ramesar 1978).

Caste erosion

In spite of the centrality of caste to Indian culture, its dissolution began on the journey to the ports and continued in the depots. Cramped circumstances on board ship made it 'virtually impossible to avoid contact with bodily emissions of various kinds', Wood (1968, 149) observed, 'a cause of the gravest defilement according to the Hindu idea of pollution'. Separate cooking arrangements at sea were provided for Hindus and Moslems, but Brahmins had to eat at a common table with lower castes and untouchables.

The system of indenture in Trinidad disregarded caste differences. East Indians were distributed among sugar estates, settled in barracks, allocated to cultivating and reaping tasks and subjected to pass laws without reference to Hindu precedence (Brereton 1979, 186). Planters' attitudes to Brahmins were variable. Some regarded them as idlers, unwilling to take on their allotted tasks; others made them *sirdars*, and used their prestige among Hindus to maintain authority on the estates (Morton 1916, 10 & 69–70).

During the latter part of the 19th century, generalisation of Indian culture obliterated many caste and regional distinctions (Naipaul 1964, 29–46). Descendants of the Calcutta emigrants – Kalkatiyas – developed a simplified Hindi derived from the Bhojpuri dialect, foregoing the niceties of grammatical expression required by caste etiquette (Durbin 1973). Shipmates (*jahaji bhai*) – even of different castes – often became closer friends than individuals who, in India, had belonged to the same caste or had come from the same district.

Trinidad-born East Indians still knew their caste labels, but no caste organisation, caste council (*panchayat*) or caste rules regulated interpersonal relations and obligations. Castes had no myths, heroes or heritage of their own; children were not initiated into caste; castes no longer disciplined their members for breaking caste rules, because there were no rules.

Enforcement of strict endogamy was undermined by the very small size and sex imbalance of many of the castes, and the requirements of the plantation régime severed the link between occupational specialisation and caste. Broken, too, was the *jajmani* system, which bound certain castes in patron–client relationships. Concepts of purity and pollution lost their significance, especially in secular contexts; untouchability disappeared, and commensality became the norm. Yet status distinctions between high and low castes survived, and caste prevailed in the restriction to Brahmins of recruitment to the priesthood and in the selection of brides and grooms according to equivalent caste status (Clarke 1967).

Family

Indians came to Trinidad in the 19th century neither as families nor as uprooted communities, but as individuals picked to fill recruiting agents' quotas. Immigrants believed – erroneously – that only men in their twenties would be accepted, and cases are recorded of a middle-aged man and a lad of 14 both claiming to be 25 – and signing indentures for Trinidad (Klass 1961, 13). Agents had the utmost difficulty in recruiting women, and the sex ratio among immigrants fluctuated continually. Planters preferred men because they could work harder, yet the shortage of women was thought to encourage immorality among the East Indians. By 1860, the government required 25% of the immigrants to be women; in 1868, the proportion was raised to 50% but then lowered to 40%; and in 1879 it was reduced still further, back to 25%. This reduction eased the recruiters' task, but family groups were often turned away in favour of single men. When the interest of the Trinidadian government shifted from temporary labour to permanent settlement at the end of the 19th century, the selection of family groups was given priority by recruiters, but it remained extremely difficult to match the supply of women to the demand (Wood 1968). East Indian males outnumbered females by three to two in Trinidad in 1891, and the ratio approached parity only at the end of World War II (Harewood 1967).

Despite the imbalanced sex ratio, East Indians maintained endogamy. Even illicit sexual relations between East Indian men and Creole women were rare. As late as 1946, East Indian–Creole mixtures numbered only 8400 in Trinidad, or just over 4% of the East Indian total. *Douglas* were much more numerous in the racially mixed towns than in East Indian rural communites, but few had Creole fathers.

During the first decades of Indian settlement the imbalanced sex ratio, lack of privacy in the barracks, and the absence of elderly people to provide models for behaviour led to domestic discord. East Indians were notorious for crimes of passion: between 1859 and 1863, 27 murders were committed, each victim the wife or mistress of the murderer (Wood 1968, 158). A visitor reported, 'a coolie regards his wife as his property, and if she is unfaithful to him he kills her without the least hesitation' (Froude 1888, 67).

East Indians preserved the essential features of the North Indian patrifocal family, though in Trinidad it contracted under economic and social pressures from an extended to a nuclear form (Schwartz 1964a, Nevadomsky 1983). Throughout the source areas from which the Kalkatiya Indians came, child marriage was the norm: nine out of ten women were married before the age of 14, and two-thirds of the men before they were 19; fathers arranged the unions, and betrothal in childhood was followed by marriage to the betrothed once the girl reached puberty (Crooke 1897, 227–8). A similar system was reported among East Indians in Trinidad in the late 19th century: 'The girls are sometimes married when four or five years of age, but they do not then go to live with their husbands' (Morton 1916, 51).

As in India, East Indian brides were expected to be virgin, and Hindu rites did not allow women who were separated or widowed to remarry. Yet, despite these elaborate in-group prescriptions, most local-born East Indians were illegitimate by Creole standards. Traditional Hindu marriages 'under the bamboo' were not legally recognised until 1946 – a decade after Moslem unions were officially sanctioned, and a century after the Indians' arrival in Trinidad (Jha 1982). A few wealthy, Westernised Hindus and Moslems did go through second, civil ceremonies, valid after 1881, to protect their children's inheritance; but as late as 1946 almost two-thirds of East Indian women were officially classified as common-law wives.

Religion

The tenacity of the Indian family was paralleled and sustained by the retention of Indian religions. Within 15 years of the beginning of indentured settlement in Trinidad, Hindu temples had been built on the estates and *saddhus*, or holy men, were travelling the country. During his visit to Trinidad in the late 19th century, James Anthony Froude saw

> a Hindoo temple, made up rudely out of boards with a verandah running round it. The doors were locked . . . so we had to content ourselves with the outside, which was gaudily and not unskillfully painted in Indian fashion. There were gods and goddesses in various attitudes . . .; Vishnu fighting with the monkey god, Vishnu with cutlas and shield, the monkey with his tail round one tree while he brandished two others, one in each hand, as clubs (Froude 1888, 66).

Here is evidence of the prominence in Trinidad Hindu folk religion of *Vishnu* or Mahadeo, the preserver, and of *Hanuman* or Mahabir, the monkey god and one of the heroes of Tulsi Dass' popular epic, the *Ramayana*.

South Indian customs and low-caste rites were also performed in 19th-century Trinidad. Madrassi fire-walking ceremonies have ceased to be held only within living memory. Goat sacrifice, presumably for a low-caste *Kali Mai* celebration, was recorded in 1849 and 1855, and in 1865 a visitor 'saw a place where the Hindus sacrifice. There was a pole with a small flag flying, a small altar of mud, and near it two stakes . . . a sort of yoke into which the neck of the goat to be sacrificed is placed and its head severed at one blow. The blood is burned on the altar and the body made a feast of' (Morton 1916, 23). During this century, however, most Hindus other than those of the lowest castes have abandoned animal sacrifice and accepted Brahminical orthodoxy.

Immigrant Brahmins rapidly established a priesthood, officiating at marriages and performing domestic *pujas* (prayer ceremonies) for their clients. In 1881, to promote *Ramanandi Panthi*, they founded the Sanathan Dharma Association, but it was not legally incorporated until 1932, when a rival Sanathan Dharma Board of Control was organised in Tunapuna. Sanskrit, though unintelligible to most East Indians, was retained as the sacred language of Hinduism until well into this century. Only with the widespread adoption of Creole English among East Indians have *pundits* (priests) used Hindi. However, the caste-exclusive nature of the priesthood, the use of a dead language, and the Brahmins' inability or unwillingness to educate the other castes has left Hinduism largely traditional and ritualistic.

Urdu-speaking Moslems were everywhere a minority in the East Indian population, yet they rapidly achieved a higher degree of cohesion than the Hindus in the 19th century. Their solidarity is perhaps explicable by their monotheism and, more importantly, by their sense of community, fostered by regular weekly worship in the mosque. Furthermore, the Moslems were less tolerant of Christianity than the Hindus, and this set them firmly apart from

Creole society and the proselytising activity of the missions. However, a common provenance in India took precedence over religious differences between Hindu and Moslems; the animosities brought by Hindu and Moslem grandfathers from India with time softened into 'a kind of folk-wisdom about the unreliability and treachery of the other side' (Naipaul, 1983, 15).

The main distinction among East Indians was not between Hindu and Moslem but between Hindu-speaking northerners who sailed from Calcutta and Tamil- and Telugu-speakers who embarked at Madras. North Indians and their descendants anathematised Madrassis, who were often darker than themselves and enmeshed in the remnants of a different caste system.

East Indians of all backgrounds acquired new traits while experiencing cultural loss. Some of the most outstanding changes were brought about by the Christian missions – Catholic in the north of Trinidad and Presbyterian in the south. The Rev. John Morton, a Canadian Presbyterian minister, settled at Iere (now Princes Town) in 1868 and moved to San Fernando three years later. East Indians were cautious about Christianity, but in 1871 two Brahmin priests were baptised, and seven years later there were 70 adult converts. Services were in Hindi, and Sunday schools instructed East Indian children in the Gospel (Morton 1916, 79–80).

The number of Presbyterians in Trinidad increased from 3400 in 1891 to 20 000, 70 years later (Samaroo 1982, 114). These gains were made largely through the conversion of Hindus, some of whom regarded Christ as an *avatar* (reincarnation) of Lord Krishna. Between 1901 and 1946 the Hindu proportion of the East Indian population dropped from 79.8% to 64.5%; in contrast, the Moslem proportion grew slowly but continually. By 1946, when Moslems accounted for 16.7% of the East Indians, they were barely outnumbered by Presbyterians and Catholics together (18.8%).

The Canadian Presbyterians, neglecting the blacks, bent their efforts to improving education among East Indians. During the 1870s missionary teachers opened schools with the financial backing of the government and the estates – indeed the Canadians were open to the charge of being too closely aligned with the colonial administration and the planters (Samaroo 1975, 1982). By the 1890s more than 50 Canadian Mission Indian Schools, as they were popularly called, served 80% of East Indian school attenders, and East Indian converts had been prepared to teach in them by the Presbyterian Training College in San Fernando (Samaroo 1975).

Hindu and Moslem boys studied at Presbyterian schools into their early teens, and many converted – those wanting social mobility, further education, and white-collar jobs, especially in teaching. By the end of World War II, three-fifths of the East Indian élite – Christian, Hindu and Moslem – were Presbyterian trained (Kirpalani *et al.* 1945, Malik 1971, 50). However, education was for boys: most East Indian girls were kept at home until they married. Despite the provision of primary schools, illiteracy – especially among Hindus – remained the hallmark of the East Indians, and as recently as 1946, 37% of men and 66% of women were illiterate in English, though some were undoubtedly literate in Hindi.

Society and space

At the end of the colonial period the separation between Creole and East Indian remained the most salient feature of Trinidad's social structure. Out of a population of 1 million in 1960, Creoles accounted for over 60%; the breakdown by colour groups was: whites 2%, browns 16% and blacks 43%. East Indians made up 37% of the island's total, among whom Hindus comprised 23%, Moslems 6% and Christians 8%.

In addition to the neighbouring island of Tobago, which is a political dependency with an almost entirely black population, Trinidad comprised five racial zones (Augelli & Taylor 1960). The western sugar belt and its subsidiary rice-growing areas (Fig. 2.2), lying between Port of Spain and San Fernando, together with the Naparimas around San Fernando and the sparsely populated south-west peninsula, were predominantly East Indian (Fig. 2.3). Port of Spain and its associated conurbation stretching along the Eastern Main Road towards Arima contained more than 250 000 inhabitants, over 90% of whom were Creole. San Fernando, with a population of 40 000, was the only other town of note. Almost three-quarters of its inhabitants were Creole, but East Indians

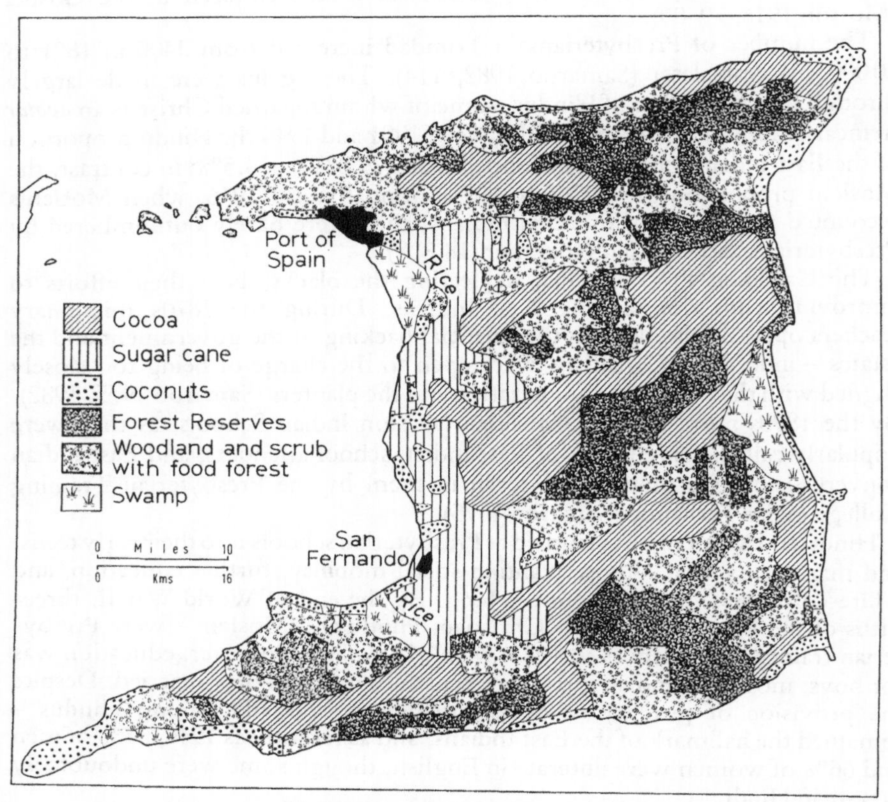

Figure 2.2 Trinidad: land use.

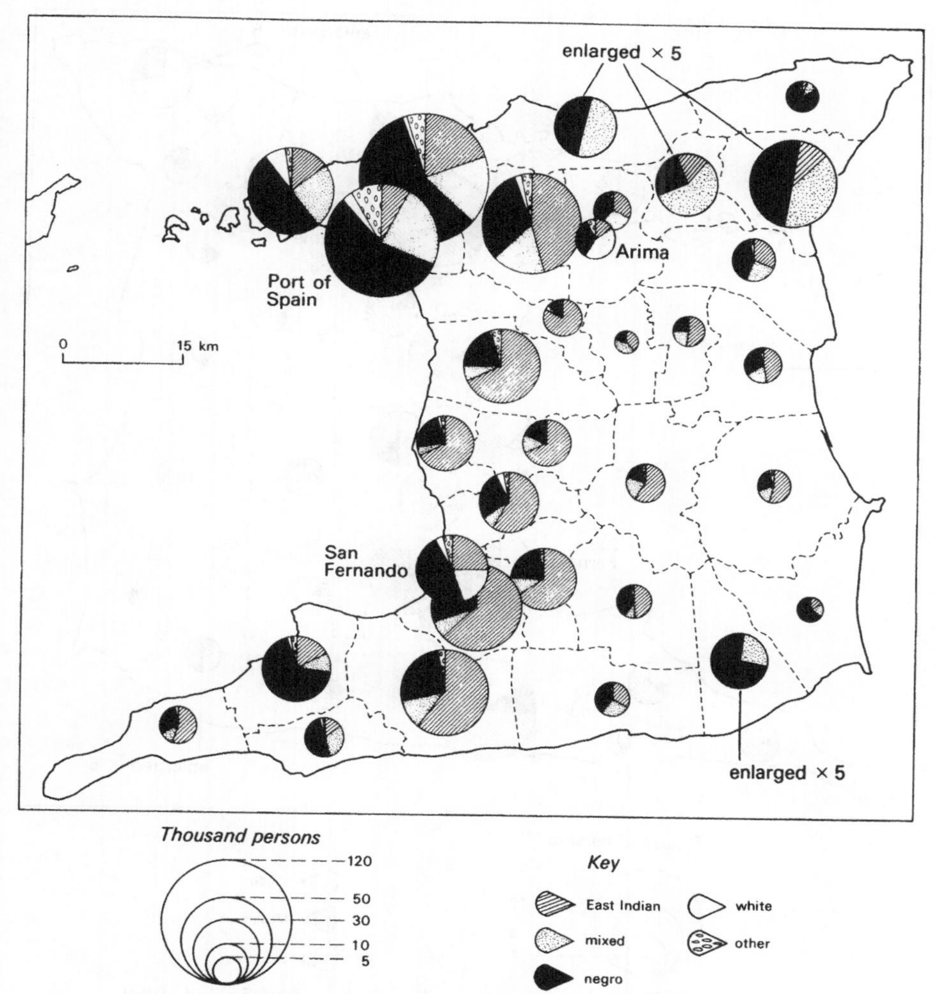

Figure 2.3 Racial groups in Trinidad, 1960.

formed a sizeable minority (25.7%) – the largest proportion that East Indians achieved in any urban settlement of note. The north and east of Trinidad were rural and sparsely populated, largely with blacks, except for Nariva where there was a pocket of East Indians. By far the most racially heterogeneous locality was the central uplands, where mixed communities of blacks and East Indians subsisted by small-scale cultivation supplemented with cocoa farming.

The outstanding feature of the spatial pattern was the contrast between town-dwelling Creoles and rural East Indians (Fig. 2.3). Whites, coloureds, blacks and Chinese were only weakly segregated from one another at the national level, and the three East Indian groups were closely aligned. Christian Indians were more urbanised than Hindus and Moslems, resided in close

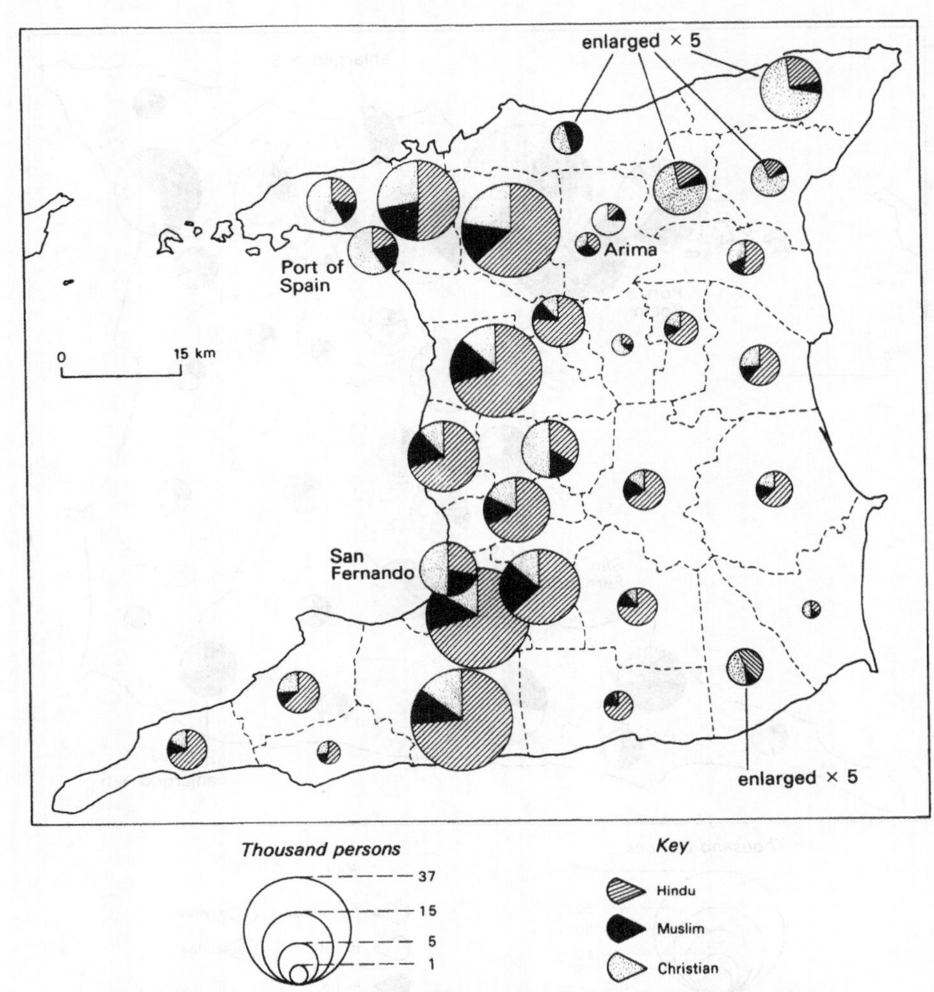

Figure 2.4 Religious groups among East Indians in Trinidad, 1960.

proximity to Creoles of mixed ancestry, but segregated themselves from blacks (Fig. 2.4) (Clarke 1976a). East Indian isolation supported a distinctive way of life and leadership. Naipaul noted:

> Living by themselves in villages, the Indians were able to have a complete community life. It was a world eaten up with jealousies and village feuds; but it was a world of its own, a community within the colonial society, without responsibility, with authority doubly and trebly removed. Loyalties were narrow: to the family, the village. This has been responsible for the village-headman type of politician the Indian favours, and explains why Indian leadership has been so deplorable, so unfitted to handle the mechanics of party and policy (Naipaul 1962, 82).

Race and politics

East Indians and blacks were separated socially and spatially, both through British design and their own mutual consent. Sharp distinctions between white overseers and black or East Indian gangs, and between East Indian field-hands and black factory workers marked 19th century sugar estates. Blacks denigrated the indentured East Indians as slaves, and scorned them as scabs who depressed wages and levels of living and thus put free labour at the mercy of the planter class. Economic rivalry between black and East Indian was reinforced by racial and cultural antipathy, and Froude, a contemporary observer, commented, 'the two races are more absolutely apart than the white and the black' (Froude 1888, 67). Wood, reflecting on that time, noted that

> the Indians, although Caucasians, were not white; although sometimes darker-skinned than many Africans, they were not negroid. They could not fit into the interplay of European and African in any predictable way. A new set of assumptions and antipathies was brought into an already complicated situation, when an alien high culture, satisfied with its own values, was imported into a society where the dominant Europeans were wanting, whether consciously or not, to fashion the Negroes in their own image (Wood 1968, 111).

East Indians and blacks in Trinidad internalised white racist stereotypes of one another. East Indians abused blacks as 'niggers' and blacks hurled back the epithet 'coolie'. 'Like monkeys pleading for evolution, each claiming to be whiter than the other', even nowadays in Naipaul's opinion, 'Indians and Negroes appeal to the unacknowledged white audience to see how much they despise one another' (Naipaul 1962, 80).

Thus when East Indians began to organise in the late 19th century they eschewed multiracial working-class alliances for racially exclusive bodies – the East Indian National Association in Princes Town and the East Indian National Congress in Couva, both Christian dominated. Political representation, too, was conceived along racial lines. The first East Indian nominee to the Legislative Council, George Fitzpatrick, who took his seat in 1912, spoke out mainly on matters of significance to the East Indian population (Tikasingh 1982, 20). So strong was East Indian fear of erosion of the Indian way of life that in 1922 they reacted to British and Creole proposals for constitutional advance with a plea – which was refused – for proportional or communal representation (Ryan 1972, Tikasingh 1982).

By the late 1920s East Indian intellectuals favoured the revival of Hindi and Urdu, and the politically minded had launched the Young Indian Party. During the depression, however, several radical East Indians, notably Hosein, Roodal and Rienzi, began to forge links with black politicians and trade unionists (Ryan 1966). They aspired to a multiracial, socialist, independent society whose ideals, for a while, were embodied in the Trinidad Labour Party. Adrian Cola Rienzi (Krishna Deonarine), the most outstanding East Indian political figure of the 1930s, analysed social problems in classical Marxist rather than racial terms – though somewhat dogmatically: 'only the dictatorship of the working class, guided by revolutionary determination and socialist philosophy can solve the contradictions and create a new social order that shall bring science and society into harmony and shall make mankind the master of the wealth that human industry and ingenuity have created' (quoted in Samaroo 1974, 87).

Electoral success for the Trinidad Labour Party was out of the question, since barely 6% of the population qualified for the vote. However, the world economic depression followed by the Italian invasion of Ethiopia fanned labour agitation and black protest, and in the late 1930s Rienzi and Roodal committed themselves to the black-led Butler Party (Jacobs 1977). In October 1936 Butler demanded better accommodation and wages for oil and sugar workers. Within a year labour riots put Butler in jail, and it was left to Rienzi, a barrister, to create the Oilfield Workers' Trade Union and the All Trinidad Sugar Estates and Factory Workers' Union – still the largest and most important labour organisations in the island (Ryan 1972).

However, during the late 1930s, ethnic groups once more polarised along political lines. East Indian areas returned East Indian candidates to the Legislature on the restricted franchise and Creoles voted for Creoles (La Guerre 1982). Rienzi was slowly absorbed into the island's legal élite and in 1944 left active politics (Singh 1982). The Creole-dominated Legislature, on the eve of the introduction of adult suffrage in 1946, had to be compelled by Britain to remove a restriction requiring East Indian voters – alone – to prove their literacy in English (Samaroo 1976). The scene was set for a new and more pernicious bout of racial politics, this time based upon 'one man one vote' and a multi-party democratic system (La Guerre 1972).

The 1946 election resulted in a tie between the multiracial Butler Party and the black radical United Front (three seats each), with two seats going to the Trades Union Council and Socialist Party and one to an Independent, but many people 'lamented the fact that race had played such a decisive role. . . . It was feared that adult suffrage had merely served to politicize and harden the cleavages in the society' (Ryan 1972, 77). The East Indian National Congress responded to the result by sending a memorandum to the British Colonial Office, protesting at their vulnerability – as a minority – to representative government, and 'Hindu pundits went up and down the country warning their flock that they would be politically and culturally swamped by the Negro majority if self-government was granted' (Ryan 1972, 83): but to no avail.

After the electoral setback of 1946, Butler's emerged as the most successful party in 1950, taking six seats, four with East Indian representatives, in a skilful return to the 1930s blend of 'sugar and oil'. However, power in the Executive Council was denied to Butler by the Governor; the East Indians moved away from him and cast in their lot with the People's Democratic Party (PDP) formed in 1953 by Bhadase Maraj, the Hindu 'strong man', head of the *Sanathan Dharma Maha Sabha*, and sugar union boss. By 1955 the PDP seemed poised for power, but the election was delayed for a year, and Dr Eric Williams' newly created People's National Movement (PNM) snatched victory by a single-seat margin with the vote in several constituencies split by numerous candidates.

Since 1956, the PNM, under mulatto and black middle-class leadership and with urban black support, has won five consecutive elections. During the period 1956–66 – in the middle of which Trinidad followed Jamaica out of the West

Indies Federation and into independence on its own (Lewis 1962) – parliamentary opposition devolved first on the PDP and, after 1958, upon the Democratic Labour Party (DLP) (Robinson 1971). The DLP was originally a coalition of opposition groups, but its voting strength depended on Hindus and, after 1960, Moslems and Christian East Indians (Bahadoorsingh, 1968).

Racial bifurcation of the electorate was convenient to Creole and East Indian political leaders (Gomes 1974). They transformed Trinidad's major segments into vote banks; whipped up racial antagonism at election periods – although stressing multiracialism at other times; and encouraged their followers to respond to their opponent's ethnic revivals by seeking and sustaining their own cultural roots. This tense political climate at the time of independence provides the context for this study of East Indians in San Fernando.

Perspectives on the East Indians

Studies of East Indians in Trinidad, carried out by American anthropologists since racial voting became so entrenched in the mid-1950s, have stressed the high degree of cultural retention and sociopolitical encapsulation achieved in rural communities. Klass, in his book on Amity near Chaguanas, emphasised that in their life cycle, kinship, caste and religion, East Indians preserved important aspects of North Indian culture. These traits set East Indians apart from Creole society. Nevertheless, Klass stressed that his village 'is part of the ongoing social, economic, and political system of Trinidad and cannot legitimately be separated from it' (Klass 1961, 248). Niehoff and Niehoff reached similar conclusions in their examination of East Indian life on the Oropuche Lagoon, south of San Fernando, and added that 'the caste system appears to be less necessary for persistence of cultural identity than the Indian family type and hereditary religious beliefs' (Niehoff & Niehoff 1960, 188).

V. S. Naipaul confirms these findings, in his penetrating, even jaundiced, interpretation – an insider's view – of East Indian life and values.

> Everything that made the Indian alien in the society gave him strength. His alienness insulated him from the black–white struggle. He was taboo-ridden as no other person on the island; he had complicated rules about food and about what was unclean. His religion gave him values which were not the white values of the rest of the community, and preserved him from self-contempt; he never lost his pride in his origins. More important than religion was his family organization, an enclosing self-sufficient world absorbed with its quarrels and jealousies, as difficult for the outsider to penetrate as for one of its members to escape. It protected and imprisoned, a static world, awaiting decay (Naipaul 1962, 81–2).

West Indian sociologists have been more cautious than foreigners in their description of East Indian behaviour as distinctive. Braithwaite, a Trinidadian Creole, in a seminal account of the island's social structure first published in 1953, confined his comments on the East Indians to the observation that 'the rise of a vocal Indian middle class and the elimination of the more demoralized type of free Indians who gave the "creole" public the stereotype of a "coolie", present a problem of major importance for the survival of the social structure' (Braithwaite 1953, 9). In 1960, however, Braithwaite noted that Creole 'values are, however, not so firmly implanted among the Indian sector (35 percent of the population) largely because of the tenacity of certain aspects of Hindu and Moslem culture' (Braithwaite 1960, 822). Nevertheless, he judged that 'although a certain amount of friction between Indian and Creole developed, and although there was a tendency to exploit the less sophisticated of the Indians, there was an over-all tolerance of the ethnic group and its culture' (Braithwaite 1960, 828).

By contrast, Crowley, an American anthropologist who carried out fieldwork in Trinidad in the 1950s, suggested that Creoles and East Indians did not form mutually exclusive categories: 'Each individual and each group has been Creolized to a greater or lesser degree in one or another aspect of his culture, while at the same time preserving inviolate the traits and complexities of his parent culture or cultures. Without losing identity, groups exchange and share members, so that relatively unacculturated individuals (rural Hindus) have a good deal of knowledge of and experience with members of groups other than their own' (Crowley 1957, 824). The crux of Crowley's argument is that 'differential acculturation and the co-existence of aspects of the Creole culture as "common denominators" between groups are the means by which the complex plural society has preserved desirable segments of each cultural entity without fragmenting the society to the point of dissolution' (Crowley 1957, 824).

The evidence reviewed earlier in this chapter shows that acculturation has taken place among East Indians, but without eradicating cultural differences – and racial exclusiveness – between them and the Creoles. As Lowenthal notes, 'purely ancestral culture and social institutions survive as ideals rather than realities or endure in name but are Creolized in character and function. Many insistent aspects of Indianness are either syncretized adaptations to Creole life or deliberate resuscitations of all but forgotten folk ways' (Lowenthal 1972, 156). Yet East Indian ways of life and thought differ substantially from Creole patterns and East Indian culture and social organisation, personality traits and values are markedly unlike those of other Trinidadians.

An extreme version of the integrative interpretation of Trinidadian society is given by the late Dr Eric Williams, the island's former Prime Minister and for 25 years leader of the Creole political party. Referring to Klass's (1961) study of Amity, he scorned the foreign student who 'may talk glibly about an Indian village in Trinidad not being West Indian. . . . The fact of the matter is, however, that in Trinidad the Negro, the Indian, French and Spaniard, English and Portuguese, Syrian and Lebanese, Chinese and Jew, all have messed out of the same pot, all are victims of the same subordination, all have been tarred with the same brush of political inferiority. Divergent customs and antipathetic attitudes have all been submerged in the common subordinate status of colonialism' (Williams 1962, 280). However, the history of Trinidad's race relations and the current involvement of race in national politics refute Williams's interpretation, and underline the need to develop social models of dominance and dissent to explain the island's racial and cultural polarity (Clarke 1984).

Theory of pluralism

These conflicting opinions about East Indian involvement in Trinidad society make it essential to establish a framework for analysing the inter-relationships among race, colour, culture and status in San Fernando. The rôle of race and colour in Caribbean societies has been keenly debated by geographers, sociologists and anthropologists. There is general agreement that the larger islands are composed of hierarchically organised strata, but some observers contend that they are based upon class distinctions, colour–class (Braithwaite 1953, Henriques 1953, Craig 1981a) or, in the case of Haiti, upon a system of colour–caste (Lobb 1940, see also Nicholls 1979). A different interpretation has been offered by M. G. Smith (1965a,b), who has stressed the importance of culture in defining social boundaries (see also Lowenthal 1972).

Smith has refined and amplified the concept of the plural society first advanced by Furnivall (1948) to facilitate analysis of the circumstances of colonialism and racial and cultural domination. Describing conditions in Burma and Java, Furnivall noted:

> the first thing that strikes the visitor is the medley of peoples – European, Chinese, Indian and native. It is in the strict sense a medley, for they mix but do not combine. Each group holds by its religion, its own culture and language, its own ideas and ways. As individuals they meet, but only in the market place, in buying and selling. There is a plural society with different sections of the community living side by side, but separately, within the same political unit. . . . Few recognise that, in fact, all members of all sections have material interests in common, but most see that on many points their material interests are opposed (Furnivall 1948, 304 & 308).

In addition to distinctions of race and culture, Furnivall pointed to other important features of the plural society: 'the union is not voluntary but is imposed by the colonial power and by the force of economic circumstances'; 'each section is an aggregate of individuals rather than a corporate or organic whole'; 'in a homogeneous society the tension is alleviated by their common citizenship, but in a plural society there is a corresponding cleavage along racial lines' (Furnivall 1948, 306, 307 & 311). Each of these observations has been elaborated upon by Smith, though he initially concentrated on aspects of cultural pluralism.

Cultural pluralism

Institutional analysis is the cardinal feature of Smith's early work on pluralism. 'I hold', he writes, 'that the core of a culture is its institutional system. Each institution involves set forms of activity, grouping, rules, ideas and values' (Smith 1965a, 79). Smith also argues that 'the institutions of a people's culture form the matrix of their social structure, simply because the institutional system defines and sanctions the persistent forms of social life' (Smith 1965a, 80).

What does Smith mean by institutions? The principal institutional systems that are involved in defining a population's culture and social relations are family, kinship, education, religion, property, economy and recreation. People who practise the same institutions and for whom they have the same values and significance form a cultural section or segment in the society. Smith clarifies his argument about cultural similarities and differences by citing the case of religion. He emphasises that 'variants of Christianity share common basic forms of organization, ritual and belief' (Smith 1965a, 84), whereas Christianity, Hinduism and Islam do not. He concludes that societies typified by minor features of differentiation but which have common basic forms of organisation are culturally heterogeneous – the term 'homogeneous' is reserved for small non-differentiated societies, whereas societies that express institutional cleavage are viewed as culturally pluralistic.

By the late 1960s, however, Smith was arguing that 'pluralism may be defined with equal cogency and precision in institutional or political terms'.

> Politically, these features have very distinctive forms and conditions, and in their most extreme state, the plural society, they constitute a polity of peculiar though variable type. The specific political features of pluralism arise in the corporate constitution of the total society. Under these conditions the basic corporate divisions within the society usually coincide with the lines of institutional cleavage (Smith 1969, 27).

Structural and social pluralism

Subsequent to his work on cultural pluralism, Smith (1966, 1969, 1974) developed the idea of structural pluralism. This occurs when population aggregates are differentially incorporated into society on a legal or political basis – for example, in the Caribbean as slaves, freemen and citizens. Smith (1969, 435) contrasts differential incorporation with uniform or universalistic incorporation, when each individual adult has full voting rights and equality before the law; and also with equivalent or segmental incorporation, where a consociation is established and the separately incorporated segments have formal parity. Differentially incorporated sections that lack group organisation are called 'corporate categories' by Smith (1966). These include assemblages of individuals such as castes, serfs and slaves. Corporate categories that develop their own internal organisation thereby become corporate groups; and in hierarchical plural societies corporate groups are usually the superordinate elements.

Smith (1969, 440) identifies three variants of pluralism – cultural, structural and social. Cultural pluralism involves institutional differences which, of themselves, do not generate corporate social difference; provided they are restricted to the private domain through universalistic incorporation, they are of personal importance but do not have structural implications for society as a whole. As Smith observes, under universalistic incorporation, 'within the limits set by law, differences of familial or religious practice are private options of equivalent status and indifference in the determination of individual civic rights' (Smith 1969, 435). Structural pluralism requires differential incorporation of the cultural sections; where it occurs, either *de facto* or *de jure*, it creates social pluralism through the projection of institutional differentiation from the private into the public domain. Structural pluralism is either based upon or creates cultural pluralism through differential access of the various sections to the society's resources. Smith concludes that 'uniform or universalistic incorporation proscribes social pluralism, though it is equally consistent with cultural uniformities or cultural pluralism among its citizens' (Smith 1969, 440). Hence, structural pluralism always involves social and cultural pluralism, and social pluralism always involves cultural pluralism; except that social pluralism may occur apart from structural pluralism when culturally distinct segments are or claim to be equal in rank and form a consociation, either *de facto* or *de jure*. This latter case occurs *de facto* in the Caribbean in Trinidad, Guyana and Surinam where East Indian–Creole contraposition gives rise to segmental pluralism.

Where cultural variations do generate corporate social differences, Smith distinguishes between the social sections and segments formed in this way and social classes. In his view, classes are 'differentiated culturally with respect to non-institutionalized behaviours, such as etiquette, standards of living, associational habits and value systems which may co-exist as alternatives on the basis of common values basic to the class-continuum' (Smith 1965a, 53). In contrast, social sections and segments possess their own value systems. They may be ranked hierarchically (sections) or may occupy parallel positions (segments) in the social order. Moreover, each section or segment may be internally stratified by class.

Relations between cultural sections are often symbiotic and maintained by patron–client links; but force or the threat of force is invariably a feature where a minority culture or corporate group is dominant. The plural model of differential incorporation demands – and supplies – an alternative interpretation to the one adopted by the consensual school of American sociology, namely that a common system of values is essential to stability in social systems and that without consensus societies cannot exist (Parsons 1952). Smith (1965a, 88) originally reserved the term 'plural society' for situations of minority cultural domination: more recently, however, he has concluded that 'plural societies are constituted and distinguished by corporate divisions that differ culturally, and that these may be aligned in differing ways to create hierarchic, segmental or complex [hierarchical *and* segmental] pluralities. It is also evident that the various modes of incorporation which are so critical for the constitution of a plural society simultaneously relate individuals and collectivities to the society's public domain' (Smith 1984a, 32).

What is a society? Smith concludes that 'only territorially distinct units having their own governmental institutions can be regarded as societies, or are in fact so regarded. Delegation of authority and governmental function is quite general and has many forms, but we do not normally treat an official structure as an independent government unless it settles all internal issues of law and order independently' (Smith 1965a, 79). It follows that even the colonial state that discharges the full range of governmental functions within its territory regulates a society distinct from that of the imperial power to which it is attached.

The pros and cons of pluralism

Critics of Smith's early work on cultural pluralism claim that it emphasises institutional differences and neglects the importance of shared values (Rubin 1960, 780–5). Others contend that it lays insufficient stress on distinctions of race and class (Wagley 1960). Another critic, referring to the problem of institutional analysis, has enquired 'at what point variations within an institutional sub-system become great enough to warrant our identification of two separate sub-systems?' (R. T. Smith 1961). However, M. G. Smith had already pointed out that 'when identical statuses and roles are defined differently we have a plurality of structural systems' (Smith 1960, 764).

Notwithstanding these reservations, it will be shown that pluralism provides an extremely useful framework for examining complex societies. For, as M. G. Smith has commented, 'pluralism itself is a multi-dimensional condition which varies in its structure, division, and intensities in different societies and in different sections or segments of the same society' (Smith 1984b, 153). The differentiating variables of pluralism have been 'relaxed' by Smith, since his early insistence on culturally incompatible institutions; they now include race as well as culture; language and differences of sect within a single universal religion; and 'secular ideologies of racism, nazism and various kinds of communism' which 'operate as primary bases for divisive incorporation of population segments in various social contexts' (Smith 1984b, 153).

Structural pluralism has something in common with Marxist perspectives on conflict, yet it avoids reducing the analytical framework to an economic dimension. Cultural pluralism emphasises the central importance of institutions, but these are broadly conceived and allow attention to be given to the currently fashionable, economic aspects of human behaviour, as well as to the racial, ethnic and cultural variables that many Marxists would dismiss as 'false consciousness' – a capitalist device to divide the working class into mutually hostile bands. In this context, it is interesting to note that both pluralists and Marxists agree in interpreting racism as a means of justifying inequality – however it is conceived.

Eugene Genovese, the historian of American slavery, is one of the few scholars to discuss the attention Marxists should give to cultural *vis-à-vis* economic factors in social relationships.

> The confusion between Marxism and economic determinism arises from the Marxian definition of classes as groups, the members of which stand in a particular relation to the mode of production. This definition is essentially 'economic' but only in the broadest sense. Broad or narrow, there is no reason for identifying the economic origins of a social class with the developing nature of that class which necessarily embraces the full range of human experience in its manifold political, social, economic and cultural manifestations. That the economic interests of a particular class will necessarily prove more important to its specific behaviour than, say, its religious values is an ahistoric and therefore an un-Marxian assumption. Since these values are conditioned only originally and broadly by the economy, and since they develop according to their own inner logic and in conflict with other such values, as well as according to social changes, an economic interpretation of religion can best serve as a first approximation and might even prove largely useless (Genovese 1971, 223).

The non-Marxist sociologist, Leo Kuper, goes further, and views ethnicity as much more potent than class when he observes, 'Ethnic sections have an origin, as basis for existence, external to and preceding the societies in which they are incorporated.' He contrasts them with classes 'that emerge only in social interaction', and concludes that, in comparison, ethnic sections are likely to have 'more enduring, comprehensive, and unique histories, a greater affinity perhaps for sentimental elaboration of identity, and a larger capacity for reasserting exclusive loyalties' (Kuper 1969, 461).

Conclusion

Pluralistic situations have many origins – conquest, forced labour, forced or free migration – Trinidad was influenced historically by all these circumstances. The utility of pluralism for understanding the island's social evolution ought to be obvious but is made clear by Smith:

> If two collectivities, B and C [Creole blacks and East Indians], are incorporated as equivalents in a single society by common subjugation to a third, A [whites], and if all members of the latter are incorporated uniformly in the public domain which it monopolizes, then all those modes of incorporation will be found within the society regulating the articulation of these collectivities. Evidently, each of these modes of incorporation [differential, equivalent or segmental and uniform or universalistic] may be transformed into either of its alternatives by altering the articulations of these collectivities, or by dissolving or crystallizing their boundaries; but such conversions of collective alignment can only proceed by political action, since this is the basis and character of all reactions between incorporated collectivities. However, the conversion of differential incorporation into a universalistic regime by political means cannot immediately eliminate those differences of culture and social organization that formerly characterized these collectivities. The dissolution of such institutional differences within and between collectivities presupposes extensive opportunities for and processes of social and cultural assimilation over a period of at least two or three generations (Smith 1974, 335).

Differential incorporation of white citizens, free people of colour and black bondsmen projected early 19th century pluralism into the Trinidad public domain, where, despite emancipation and the vote, it remained at independence. The East Indian segment, introduced under a system of indentureship, was differentially incorporated – legally as unfree labour for five years, and politically as non-voters until 1946. However, Indian culture, including religion, the family and caste, though simplified, continued substantially intact; indeed, Hindi and Hinduism experienced a revival after World War II, partly as a result of political competition between East Indians and Creoles.

> To be an Indian from Trinidad, then, is to be unlikely and exotic. It is also to be a little fraudulent. But so all immigrants become. . . . Immigrants are people on their own. They cannot be judged by the standard of their older culture. Culture is like language, ever developing. There is no right and

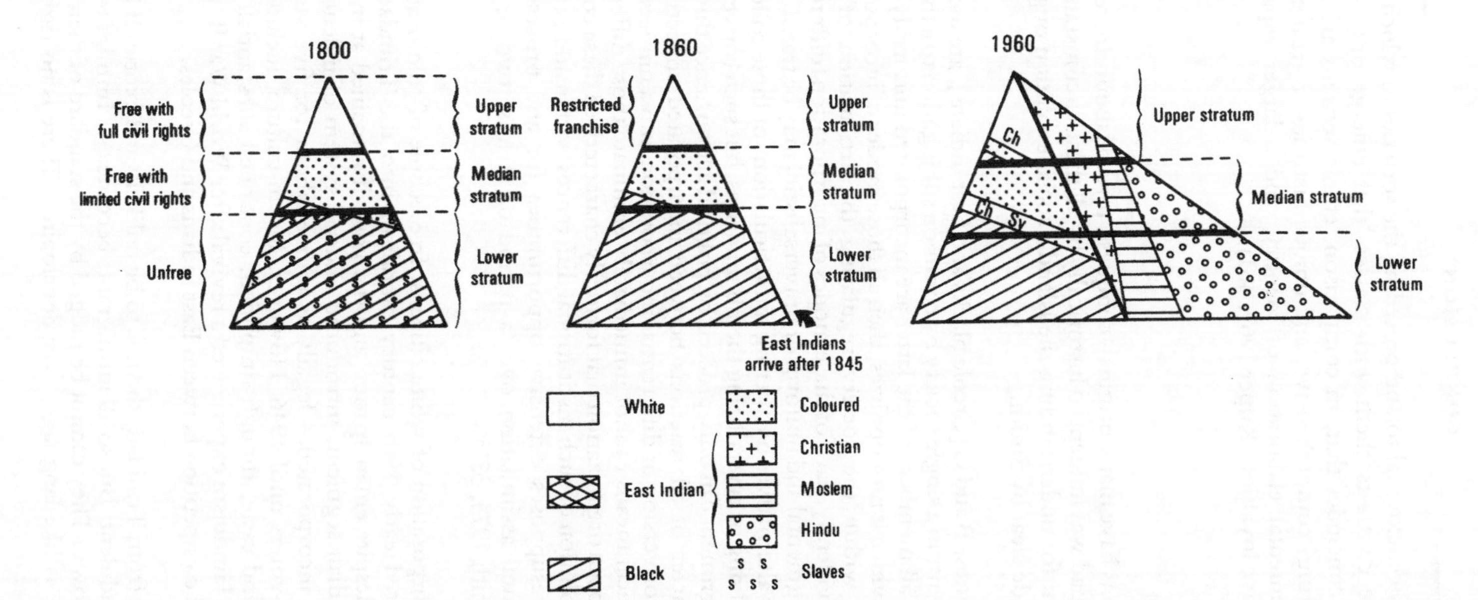

Figure 2.5 Plural framework of Trinidadian society, 1800–1960.

> wrong, no purity from which there is decline. Usage sanctions everything. And these Indians from Trinidad, despite their temples and rituals, so startling to visitors, belong to the New World (Naipaul 1972, 35–6).

East Indians formed an endogamous segment, or more accurately a series of religiously defined endogamous segments, external to the Creole stratification, with Christian East Indians closer to Creole behaviour and values than Hindus and Moslems (Fig. 2.5). This complex plurality, combining hierarchical and segmental patterns, was expressed in urban–rural (Creole–East Indian) distinctions. The parallel ranking of Creoles and East Indians at independence must not, however, obscure the fact that until World War II Trinidad was a plural society whose large and divided non-white segments were differentially incorporated and dominated by whites. For almost 150 years, circumstances of structural, social and cultural pluralism produced a highly structured system that could be broken only by rebellion or by uniform incorporation on the basis of democracy. Only when Britain began the long process of constitutional decolonisation by introducing adult suffrage in the mid-1940s did the ballot box, at last, replace coercion by the military and the police as the final arbiter.

However, universalistic incorporation at law could not abolish the empirical reality of co-ordinate social pluralism. The two major co-ordinate segments in Trinidad regrouped to compete for political power at the elections, and political organisation transformed these corporate categories – East Indian and brown and black Creole – into corporate groups. The major electoral polarisation was between East Indian and brown and black; but Trinidadian politics in the late 1950s also involved a struggle between brown (and middle-class black) Creoles and Brahminical Hindus to fill the vacuum created by decolonisation, with the Creoles attempting to woo the Moslems and Christian East Indians.

These observations apply to Trinidad as a whole, but what was happening in San Fernando during the colonial period? More specifically, how did the East Indians fare in an urban environment, and what kind of town was San Fernando? To answer these questions we turn to a consideration of the social history of the town, before making specific reference to its pluralistic structure in the decades from 1930 to 1960.

3 *San Fernando: from slavery to independence*

San Fernando lies on the Gulf of Paria some 56 km south of Port of Spain (Fig. 3.1). The town comprises a higgledy-piggledy mixture of wooden and concrete structures, with ochre-painted, corrugated-iron roofs. It was described by de Verteuil in 1884:

> It is partly built in a sort of small recess, formed by two spurs stretching from [Naparima Hill] towards the sea, and partly round the basis of the hill. Houses, generally poor and miserable, are scattered over this site of the town, except, however, along the principal street, which leads from the wharf to the foot of the hill; it has a rather winding direction. To the southward is the high ground, upon which stand the Roman Catholic church, the hospital, the town-hall and the court-house. An alley planted with trees forms a promenade; the promenade was projected by Lord Harris, and bears his name. This part is the most pleasant section of the town, as it commands an extensive view of the harbour and adjacent country (de Verteuil 1884, 303).

This description was largely echoed in my wife's entry in our diary in 1964:

> Houses are perched one on top of the other on the slopes leading down to the sea. Most are on wooden, stone or concrete stilts – almost all look dilapidated and frail. The shopping centre is untidy, the gutters littered with garbage, banana skins and orange-peel. Pungent aromas pour out of gloomy food shops. Poorly dressed men and women squat on the pavement trying to sell their wares – corn on the cob, fruit, vegetables, and jewellery. Compared to Port of Spain, San Fernando is dusty and old fashioned – but it is not without distinct character, and there are lots of East Indians, particularly in business and on the buses.

The original settlement called San Fernando, located to the north of the High Street, was developed to serve the sugar estates in the Naparimas established by French-speaking Catholics who were given permission by the Spanish Crown to settle in Trinidad in 1783. In 1784 the Governor of Trinidad, Don José Maria Chacon, celebrated the birth of the royal heir to the King of Spain by naming the settlement San Fernando de Naparima. However, the French called the town Petit Bourg, and this remained its informal name into the 20th century.

The land now occupied by the town centre was granted to Isidore Vialva and sold by him to Jean Jaillet. Jaillet carved out of this property a small estate called Mon Chagrin, and illegally disposed of the remainder as building lots. These

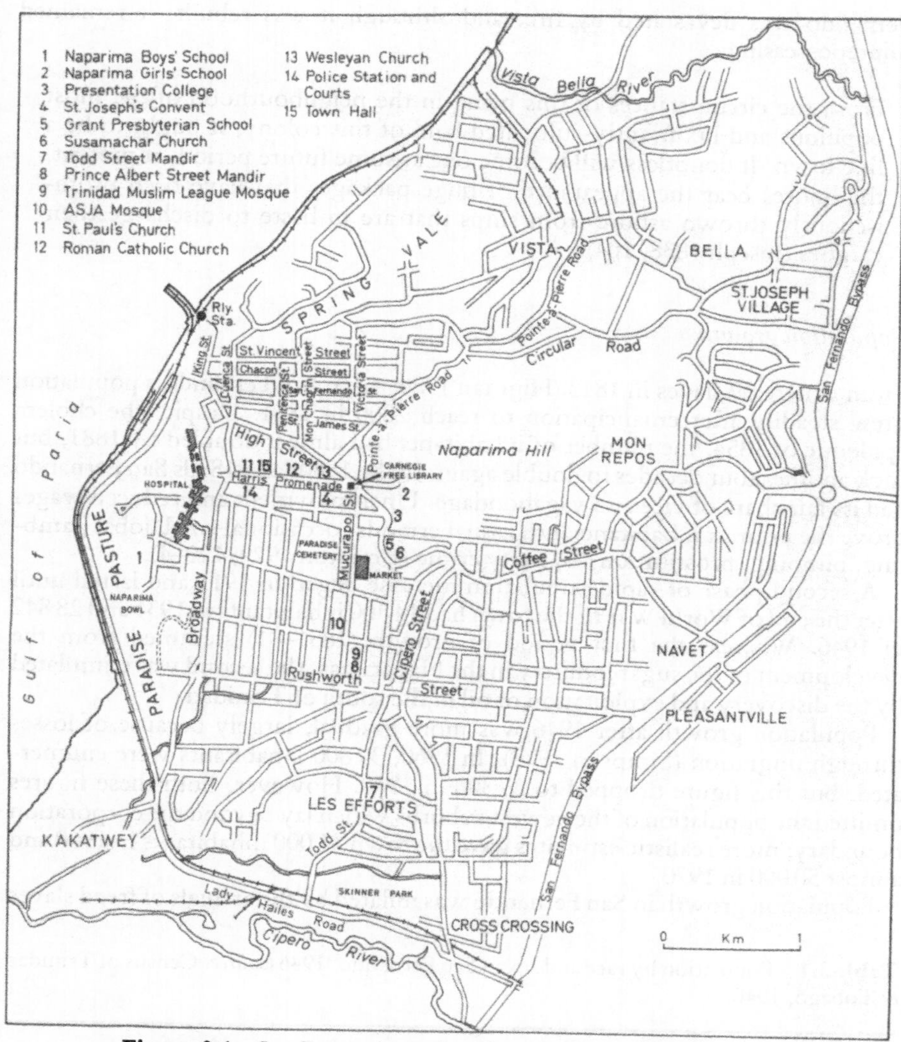

Figure 3.1 San Fernando: major place names and institutions.

illegal contracts were condoned by the Spanish administration and remained uncontested by the British after the capitulation of the island in 1797. St Vincent, Penitence and Quenca Streets were laid out on the southern slopes of the Spring Vale ridge. A market place occupied the small central square between St Vincent and Chacon Streets, and a rest house for visitors was situated near the harbour at the corner of St Vincent and King Streets (Ottley 1971).

San Fernando's development flagged during the early 19th century. The sugar industry of the Naparimas lacked capital, slaves and the political security which would come only with formal British annexation in 1815. Three years later San

Fernando was devastated by fire, and although it was rebuilt, it remained unprepossessing:

> From the circumstances of this being in the neighbourhood of the most populous and luxuriantly-cultivated land of this colony, it ought to be a fine town. It doubtless will become one at some future period; at present, the houses bear the appearance of huge packages that have been promiscuously thrown ashore from ships that are in haste to discharge their cargoes (Joseph 1838, 104).

Population growth

From a mere 99 slaves in 1813 (Higman 1984, 427), San Fernando's population grew steadily after emancipation to reach 2800 in 1851. Despite the cholera epidemic of 1854, the number of inhabitants had almost doubled by 1881, but took another four decades to double again. In the 1880s and 1890s San Fernando had its fair share of vice and vagabondage. Unemployment and reduced wages drove the poorest inhabitants into casual errands, occasional small jobs, gambling, pimping, prostitution and petty crime (Brereton 1979, 123–4).

A second phase of rapid population increase began in 1911 and lasted until after the end of World War II; the town had 14 400 inhabitants in 1931 and 28 842 in 1946. Whereas the first period of growth (1851–81) stemmed from the development of the sugar industry in the Naparimas, the second was stimulated by the discovery and exploitation of oil in the south of Trinidad.

Population growth after 1946 was more gradual, largely because of losses through migration (Simpson 1973). In 1960, 39 800 inhabitants were enumerated, but this figure dropped to 37 300 in 1970. However, both these figures omitted the population of the newest suburbs which lay beyond the corporation boundary; more realistic estimates gave the town 41 000 inhabitants in 1960 and almost 50 000 in 1970.

Population growth in San Fernando was inflated by the arrivals of freed slaves

Table 3.1 Population by race and sex in San Fernando, 1946 (*source:* Census of Trinidad & Tobago, 1946).

Group	Sex				Sex ratio
	Male	Female	Total	%	(M per 1000 F)
white	553	479	1032	3.58	1154
black	7313	8452	15 765	54.66	865
East Indian	3480	3326	6806	23.60	1046
Syrian	88	62	150	0.52	1419
Chinese	346	122	468	1.62	2836
mixed or coloured	2077	2529	4606	15.97	821
not specified (including Caribs)	5	10	15	0.05	500
Total	13 862	14 980	28 842	100.00	925

after 1838, of time-expired Indians in the 1850s and 1860s, and of blacks – especially from the Windward Islands, Antigua and Barbados (Johnson 1973) – and East Indians during the establishment of the oil industry. Slack employment periods witnessed the influx of domestic servants because of the withdrawal of women from agriculture; booms brought in male labourers. The sex ratio dropped from 1020 males per 1000 females in 1871 to 879 in 1901 and 869 in 1931, but recovered thereafter to 925 in 1946 and 1960. By the end of World War II sex ratios among the various groups had bifurcated in a striking way. Among black and coloured Creoles women predominated, but in all the immigrant minorities – white, East Indian, Syrian and Chinese – a superfluity of men was recorded, the sex ratio ranging from 1046 males per 1000 females among East Indians up to 2836 for Chinese (Table 3.1).

Race and status

During the first three decades of the 19th century the social structure of the small settlement of San Fernando was characterised by racial stratification and underpinned by legal distinctions. White masters occupied the narrow apex of the social pyramid; black slaves the broad base. Free coloureds, the descendants of white masters and slave women, were allocated an intermediate position in the social hierarchy.

The social system was less coercive, but the social structure more complex, in Trinidad in 1800 than in other British West Indian islands. Plantation slavery was only just becoming established in Trinidad; English-speaking whites were outnumbered by Spanish colonists and French émigrés from Saint Domingue; and free coloureds – French, English or Spanish in speech – were more numerous than all the whites put together.

After emancipation in 1834 estate owners and managing attorneys set up home alongside the merchants in San Fernando. So, too, did coloured clerks and storekeepers, black artisans, domestics and labourers. The town was rapidly developed as a Creole community in which colour and occupation denoted a person's social standing. Brereton (1979, 97 & 203–4) cites two blatant cases of racial discrimination and intolerance in San Fernando in the 1870s, in which the injured party was brown or black. Moreover, in 1867, the working-class inhabitants of San Fernando, 'descendants of Africa', petitioned against municipal taxes and general oppression by the Borough Council. Some ascribed their lack of employment to the deliberate flooding of the colony's labour market by indentured Indian immigration (Brereton 1979, 148–9). Of course, the sugar estates in the town's hinterland became exclusively East Indian in population during the course of the 19th century.

Gradually, East Indians who had completed their indentures began to take up residence in San Fernando. In 1861 the Indian-born population numbered 195, or less than 5% of the town's inhabitants. Ten years later Indian immigrants totalled 414, local-born East Indians 132; together they accounted for 11% of the population. Many East Indians worked as milk vendors, gardening labourers, scavengers and market sellers; a small number invested in retailing and property (Singh 1974, 47). In 1873 five East Indian shopkeepers from San Fernando bought the Corial estate for $18 000 (Trinidad & Tobago), hired a manager, and themselves applied for an allocation of indentured labourers (Wood 1968, 277).

The East Indian trading community in San Fernando was augmented by the arrival of a small number of Bombay businessmen after World War I. The Bombay Indians kept socially aloof from the descendants of the indentured immigrants, but between them these two groups eventually owned the majority of dry-goods shops on the north side of High Street (Plate 1). The Indian-born, including ex-indentured servants, numbered only 262 in 1931, when the entire East Indian population accounted for little more than 17% of the town's population. Nevertheless, one visitor, almost a decade earlier, had erroneously claimed that San Fernando's 'population is so overwhelmingly East Indian that even the English are forced to learn Hindustanee'. More accurately, he went on, 'His worship the mayor is a Hindu; on certain days of the week the visitor who strolls through its wide asphalted streets might easily fancy himself in a market city of Central India, ... [with] such signs as "Sultan Khan, Pawn Broker", "Samaroo, Barber" or "Jagai, Licensed to deal in cocoa. . . ." (Franck 1923, 398).

Other minority groups also throve on commerce. Syrians penetrated the dry-goods trade during the boom years of the 1920s, and Chinese male immigrants rapidly established themselves in groceries, laundries and restaurants. By 1946 the Chinese and Syrian minorities filled status gaps in the Creole social hierarchy and accounted for 1.6% and 0.5% of San Fernando's population, respectively. East Indians (23.6%) remained outside the Creole stratification of whites (3.6%), coloureds (16.0%) and blacks (54.7%) (Table 3.1).

The development of the oil industry after 1910 brought white technicians and administrators into the area, though senior staff were mostly housed on the company's compound at Pointe-à-Pierre. Skilled jobs commanding good wages became available to a generation of blacks and East Indians in San Fernando. Some invested their money in homes; others bought cars and augmented their incomes as part-time taxi-drivers. Nevertheless, as late as the 1960s only one major shop on High Street was Negro-owned.

The best-paid and most prestigious occupations remained white preserves. Senior posts in the town hall, managerial positions in banks and branches of Port-of-Spain commercial houses, and ownership of the largest businesses and stores were largely, if not exclusively, in white hands until well after World War II.

Whites were socially distant from all the other racial groups, but aloofness also featured in the potentially more competitive relationship between East Indians and black or coloured Creoles (Samaroo 1975). Blacks stereotyped the 'coolies' as parsimonious, rustic and isolated, and complained about their use of family links and racial solidarity in business deals and political alignments. East Indians in turn scorned what they saw as the loose behaviour of the 'niggers', their promiscuity, their failure to exercise parental responsibility, their lack of financial forethought and their racial features. Each group treated the other as socially beneath itself; each appealed to white norms in an attempt to prove its own greater proximity to European culture – the blacks by emulating white institutions, behaviour and attitudes; the East Indians by invoking their Aryan ancestry, long cultural history, supposed morality and racial purity.

Religion

From its beginnings during slavery, San Fernando was essentially a Catholic town, and a Roman Catholic church was built in 1823. However, soon after, the Anglicans completed their own place of worship, and despite white hostility, the Methodists opened a ministry for the town's free Negroes and coloured people.

After emancipation the British embarked upon a policy of Anglicisation of the law and religion (Wood 1968). French- and Spanish-speaking Roman Catholics – white, brown and black – resented and resisted these changes. In San Fernando the French and Spanish languages were soon replaced by English, but the Catholic Church survived: by 1851, 60% of the town's population was Catholic and only 25% Church of England. Finally, during the 1860s, a compromise was worked out which placed the Catholic Church on an equal footing with the Anglican.

Neither major denomination carried out missionary work among the East Indians in the south of Trinidad. However, the Presbyterian Church, under the guidance of the Rev. Morton, brought East Indians into contact with Christianity, and gradually into residence in San Fernando (Morton 1916). The first Canadian mission school was started in an old building on Cipero Street in 1871 and moved to the site of what is now the Grant Memorial School in 1872. In the same year Susamachar – the Presbyterian Church of Good Tidings – was opened on a lot adjacent to the school (Fig. 3.1). East Indian cultural distinctiveness also came to be expressed in the townscape of the southern neighbourhoods. A mosque was built between Rushworth Street and Harris Promenade around World War I, and Hindu temples were eventually added during the vigorous building programme initiated by the Sanathan Dharma Maha Sabha after World War II (Fig. 3.1).

By the end of World War II there were 6806 East Indians in San Fernando, of whom 2750 were Presbyterian, 1790 Hindu and 1759 Moslem. Hindus and Moslems were despised by the Creoles because they were heathen and clung to traditional customs, costumes and modes of address. In contrast, Presbyterians were more Westernised in their behaviour and usually better educated. Consequently, they were more readily accepted by the Creoles and better able to secure some of the prestigious white-collar jobs, which, in the south of Trinidad, were available only in San Fernando. 'Only the urban Indian, the Indian of the middle class, and the Christian convert were able to move easily out of the Indian framework', Naipaul (1962, 82) noted. 'The Indian Christian was more liberal and adaptable in every way; but, following far behind the Negro on the weary road to whiteness, he was more insecure.'

Town plan

The morphology of central San Fernando was fashioned in the middle of the 19th century. At that time there were 463 houses, and the built-up area was confined to the Spanish colonial grid with extensions along San Fernando and St James Streets, High Street and the major arterial route-ways. Unaligned with the east–west town grid, High Street ran along the bottom of the Mariquire Ravine; it was the main thoroughfare leading to the harbour and contained the commercial centre of the town (Fig. 3.1).

Harris Promenade, a gift from the government, occupied the ridge to the south of High Street and in the 1850s became the administrative and ecclesiastical centre. Immediately to the south, Paradise Cemetery was sited on the limits of the town. Suburbs were coming into being beyond the compact built-up area. Scattered settlement followed Pointe-à-Pierre Road and Circular Road to the north, Coffee Street, winding eastwards round the southern side of Naparima Hill; and Cipero Street, which led southwards off Coffee Street into the canefields.

During the following 50 years, town growth was very gradual and confined to a broad-meshed grid across the flat land between Harris Promenade and Rushworth Street, and between Cipero Street and Broadway. Broadway followed the alignment of the proposed extension of the sugar tramway linking the Cipero *embarcadère* with the San Fernando wharf. The tramway was never constructed, but a cutting made for it in the 1860s severed Harris Promenade from the open space at Paradise Pasture (Collens 1888, 147).

Spring Vale ridge was colonised after 1900 and the town expanded inland to the north and south around Naparima Hill, incorporating dispersed rural dwellings and, later, the nucleated settlement of St Joseph's Village. The most densely built-up of these early 20th century suburbs lay to the east of Cipero Street, between Coffee Street and Rushworth Street. Buses and taxis served the town in the 1920s, and by 1950 the route system extended from the Cipero River in the south to the Vista Bella River in the north.

After 1950, government housing was located at Mon Repos, Navet and Pleasantville, on the site of abandoned sugar plantations lying to the east of the original town. Private housing developments occupied former cane land at Les Efforts to the south of Rushworth Street and at Marabella, north of the Vista Bella River, on the road to Pointe-à-Pierre.

Non-residential land use

Four major elements of non-residential land use were in evidence by 1900, and deserve re-emphasising. There was the port and warehousing area built near the wharf; the railway station adjacent to it; the commercial axis formed by the High Street running inland from the harbour; and the public buildings erected on the ridge occupied by Harris Promenade.

Twentieth century building added to, but did not displace, these patterns. The retailing district extended inland along Coffee Street and down Mucurapo Street to the site of the corporation market. Banks were established, and the large multiple stores on the south side of High Street included a Woolworth's and branches of Port-of-Spain firms. A shopping precinct with a supermarket, several boutiques and car-parking facilities was opened at the Carlton Centre between High Street and St James Street in the mid-1960s (Giacottino 1977).

At independence Harris Promenade remained the administrative and ecclesiastical centre. The Anglican and Roman Catholic churches, the town hall, several schools, the police station, the law courts, and many barristers' and lawyers' offices were located there. A narrow strip of park ran the length of the promenade and separated the traffic flows. The park had statues of Christopher Columbus – the island's discoverer – and Mahatma Ghandi, whose memorial was subscribed by local East Indians and Bombay businessmen, and provided recreational space for townspeople, for those attending court, and for those who came to watch the public television screens, mounted on stands opposite the town hall.

Recent decades have witnessed major improvements in town recreation and health provision. A public library was constructed at the junction of Harris Promenade, High Street, Pointe-à-Pierre Road and Coffee Street – Library Corner – in 1919, and several cinemas opened in the 1930s. Skinner Park, a portion of the old Les Efforts sugar estate, was presented to the town as a recreational centre by Usine Ste Madeleine in 1930, and in 1962, the government built a theatre and arena at the Naparima Bowl in Paradise Pasture. The new hospital, also in Paradise Pasture, opened in 1955.

Population and housing

Population distribution and density in the first third of the 20th century fitted the mould established prior to 1900 (Fig. 3.2). Population density in 1931 reached more than 35 persons per acre around Chacon Street, and rose to over 51 in the Coffee Street–Cipero Street area. The residential neighbourhood between Cipero Street and Broadway had already been peopled, and housing had expanded along the Pointe-à-Pierre Road to the north of Naparima Hill, but the coastlands and the town's inland periphery remained sparsely inhabited.

By 1946 the inhabitants of San Fernando remained, as in 1931, concentrated, but generally at higher densities, in those parts of the old town which had been laid out prior to 1850 or in the late 19th-century suburbs located immediately to the north of Rushworth Street. All these districts averaged more than 48 persons per acre, and at the junction of Coffee and Cipero Streets the density exceeded 66 persons per acre. Moderate densities ranging between 12 and 33 persons per acre were recorded in the newer areas wrapped around Naparima Hill, in the commercial centre along High Street, and in the neighbourhoods extending down to the port. However, the northern and southern periphery and the coastal belt, too, were almost as sparsely occupied as in 1931.

San Fernando's 28 800 inhabitants in 1946 occupied 6800 houses at a density of 1.6 persons per room. Sufficient accommodation had been built to reduce the ratio found in 1931, but almost two-fifths of the dwellings comprised only one room, and 80% were wooden. About 70% of homes were tenanted. Only 65% had water closets, and there were 240 outside faucets to serve those houses that lacked a piped supply of water.

Despite 1950s slum clearance and rehousing in the areas adjacent to Cipero Street, population densities in 1960 remained at over 40 persons per acre in the established districts in the south of the town. The commerical centre lost population whereas the northern and southern peripheries gained inhabitants, recording densities of more than 10 persons per acre. Nevertheless, a sparse

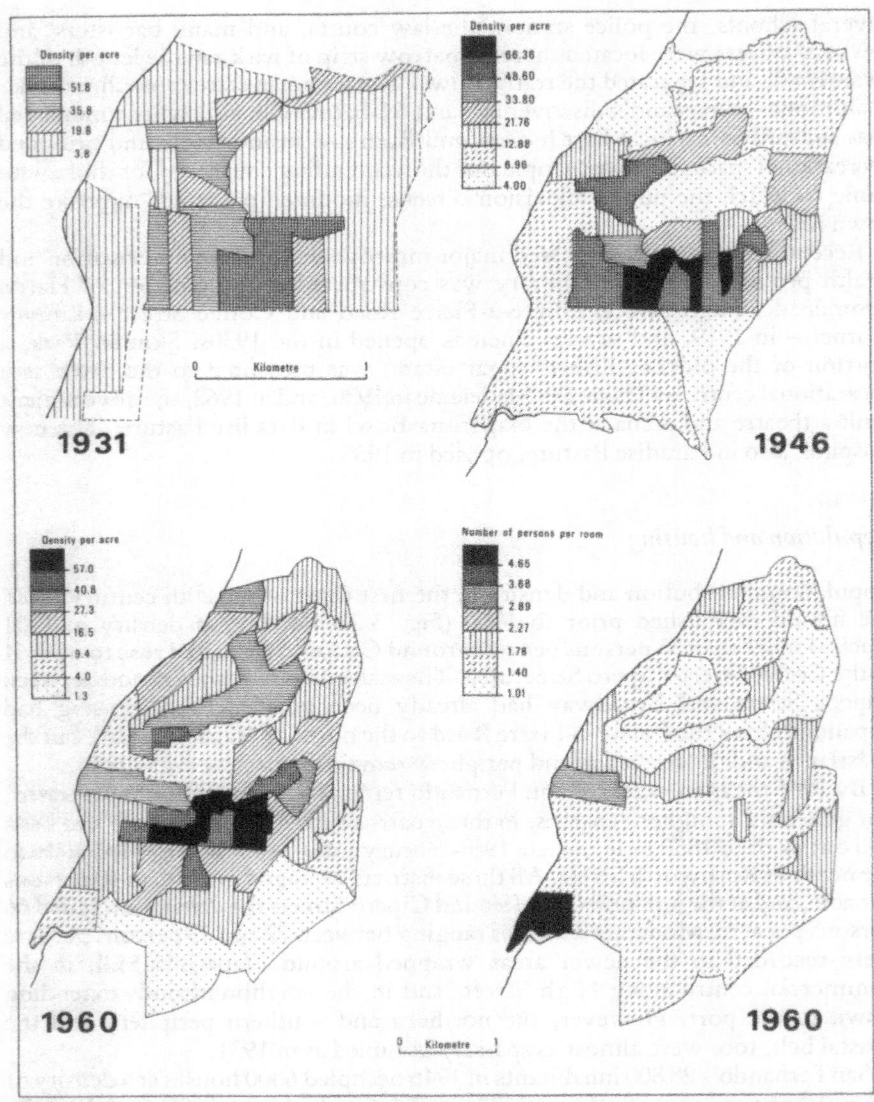

Figure 3.2 Distribution and density of population in San Fernando, 1931, 1946 and 1960.

population characterised areas as diverse as St Joseph's Village, the government housing projects near the bypass, and the squatter settlement at Kakatwey on the corporation garbage tip lying adjacent to the mouth of the Cipero River.

Most sections of the town achieved ratios of fewer than 1.4 persons per room in 1960. The lowest ratios were recorded at St Joseph's Village and extended west along Circular Road into the old town and up the slope into Spring Vale.

Densities greater than two persons per room typified older neighbourhoods close to the town centre and on the north side of Naparima Hill, and new districts such as Pleasantville, the hospital and Kakatwey. At Kakatwey each single-room shanty housed more than four occupants.

Economy

The development of San Fernando during the early 19th century was intimately associated with the growth of the sugar industry in the Naparimas, where Scots and English planters soon outnumbered the French. Small sugar estates were amalgamated to form larger and more viable units. Ste Madeleine, a consolidated property of the Colonial Sugar Company, about 3.2 km east of the town, was for many years the largest central sugar factory in the British Empire.

In 1868 San Fernando was said to look 'as though a space just big enough had been cleared from the sugar cane that well-nigh surrounded it. We found it to be, however, a place of some importance with a brisk trade especially in the crop season, being the market town and port for the largest sugar district in the island' (Morton 1916, 21). A flourishing community of Scottish merchants dominated commercial activity; during the 1870s Burns's birthday was celebrated with 'Scottish reels and songs, haggis, herrings and Scotch' (Brereton 1979, 52).

The town's commercial and administrative functions were enhanced by its communications system. A ferry service provided an important link between the south of the island and Port of Spain between 1818 and its discontinuance in 1931. Road links were poor until the government initiated a substantial programme of building in the 1860s; a railway service from Port of Spain operated from the 1880s to the 1960s. Since the early 1930s San Fernando has been the hub of a bus and shared-taxi service which covers much of southern Trinidad, and, more recently via the new highway to the north, has fed into networks radiating from Curepe and Port of Spain.

San Fernando's growth and prosperity during this century, however, have been due principally to the oil industry. Oil was struck at Guayaguayare on the south-east tip of the island in 1902. By 1910, 25 oil companies were prospecting, and a year later San Fernando – the site of the nearest harbour – exported its first cargo of crude oil. In 1913 Trinidad Leasehold purchased the abandoned sugar estate at Pointe-à-Pierre, on the northern outskirts of San Fernando, for an oil refinery. Three years later crude oil was being pumped north to Pointe-à-Pierre from wells at Forest Reserve, and the refinery was manufacturing gasoline, fuel oil, kerosine and gas oil for British use in World War I. Oil production soared from 125 000 barrels in 1910 to 2 million in 1920 (Ottley 1971, 115).

Development continued at Pointe-à-Pierre throughout the interwar years, even when other Trinidadian affairs were severely depressed. A petroleum cracking plant was built in 1927 and three more refining installations were added between 1930 and 1938. A plant producing low-octane aeronautical fuel was completed in 1938 and during World War II the British government helped Trinidad Leasehold construct another refinery, solely for aircraft fuel, alongside the original one. By 1956, when Texaco bought out Trinidad Leasehold, one commentator claimed that 50% of San Fernando's labour force worked in the oil industry in one capacity or another, and 30% were employed at Pointe-à-Pierre alone (Ottley 1971, 115). These figures probably overestimate the case, but oil refining did give San Fernando a degree of wellbeing rare among West Indian towns at this period: in 1946, for example, fewer than 3% of the town's labour force were unemployed. Moreover, the oil industry and its supporting services provided opportunities for economic mobility among male non-whites, though women remained largely trapped in personal or domestic service (Table 3.2).

Education

Education in San Fernando was neglected throughout the first half of the 19th century; only in 1859 did a borough school open. Its headmaster in the 1880s was J. J. Thomas, a black Trinidadian famous for his study of Creole grammar and his vigorous rebuttal of Froude's characterisation of the coloured people of the West Indies as inferior to whites (Thomas 1969). The church's rôle in primary education was not settled until the 1860s; even so, for more than half a century thereafter, the Anglican and Catholic schools had to contend with the planters' charge that education spoilt labour.

However, the churches did secure major secondary school initiatives. St Joseph's Convent was established in 1882 and re-opened in 1935. St Benedict's College was founded in 1935 and later renamed Presentation College. Not to be outshone by the Catholics, the Presbyterian missionaries affiliated with Queen's Royal College in Port of Spain in 1900 to allow boys at Naparima College to sit public examinations. In 1917 Naparima Girls' School was founded (Fig. 3.1).

All these schools were fee paying, but they enabled a middle class, created by the oil boom, to educate their children for the professions and the civil service. These schools also provided avenues for advancement for a few scholarship winners of humble background. Thus, by equipping students for white-collar posts, the Presbyterians provided at least a few East Indians with an alternative to the canefields and an opportunity to escape from 'coolie' poverty.

By 1946 illiteracy in San Fernando in general was down to 8%. Educational provision accelerated after World War II, and from the late 1950s students were admitted to government-approved secondary schools by scholarship only. Schoolchildren were drawn into San Fernando from all over the south of Trinidad; many even boarded in the town. High standards were quickly achieved, and in 1970 pupils attending San Fernando's secondary schools won eight out of ten island scholarships for university study.

Politics

The more numerous Creoles have dominated political affairs in San Fernando throughout its history. Whites surrendered their control of the town council only when universal adult suffrage was introduced in 1946. Despite overall white domination, individual East Indians achieved prominence after World War I. An East Indian mayor was elected in San Fernando in 1919, and during

Table 3.2 Occupations in San Fernando, 1946 (*source:* Census of Trinidad & Tobago, 1946).

	Male		Female	
	Total	%	Total	%
agriculture	214	2.68	28	0.83
minerals and quarrying	442	5.53	14	0.42
fishing	92	1.15	0	0.00
forestry	18	0.23	0	0.00
manufacturing and repair	2552	31.97	807	23.97
construction	1157	14.49	7	0.20
transportation	678	8.50	57	1.69
commerce	1215	15.22	559	16.60
recreation services	52	0.65	13	0.39
professional services	282	3.53	437	12.98
public services	649	8.12	48	1.43
personal services	405	5.07	1364	40.51
ill-defined or not specified	227	2.84	33	0.98
Total	7983	100.00	3367	100.00

the late 1930s and early 1940s Rienzi, Roodal and Ramsarran succeeded one another in that office.

Rienzi and Roodal were both involved with the oil industry. Roodal, a well-known politician of Madrassi origin, owned extensive tracts of oil land in the south. Rienzi was a prominent barrister and agitator among the oil workers, who, under the messianic leadership of Tubal Uriah Butler, carried out a number of crippling strikes in 1937 in protest against low wages. The strikes closed down San Fernando, started a riot, and marked the beginning of decolonisation in Trinidad. The labour movement in the late 1930s involved substantial interracial solidarity, particularly among sugar and oil workers, and suggested that blacks and East Indians might co-operate politically to form a working-man's party once adult suffrage was introduced.

After the introduction of the adult vote in 1946, racial co-operation in national politics between blacks and East Indians was rarely achieved and never sustained. San Fernando proved an exception. Its public life for the decade after 1946 was dominated by Roy Joseph, a creolised Syrian–Indian, whose political machine, known as the 'eighth army', drew its support from the Catholic (black) and Moslem (East Indian) groups (Ryan 1972, 147).

The alliances were broken in 1956 when Dr Eric Williams formed the PNM, with considerable support from the Creole middle class in Port of Spain and San Fernando. Hoping that a platform of socialism and anti-colonialism would bind together all non-white sections of the society, he successfully drew Moslem and Christian East Indian voters away from the Hindu-led PDP and secured their support in the crucial urban marginals (Malik 1971, 38–42). By this manoeuvre the two newly created constituencies in San Fernando fell to the PNM, and in one of them Roy Joseph was soundly beaten by Dr Winston Mahabir, his Christian

East Indian opponent – a crucial factor in the PNM's one-seat victory in the general election. A similar constellation of racial and religious factors influenced the outcome of the 1961 poll. Gerard Montano, son of a wealthy, near-white San Fernando family, retained his San Fernando East seat for the PNM, and Said Mohammed replaced Mahabir (who resigned from the PNM and emigrated to Canada) as the East Indian (Moslem) vote-catcher – and elected representative – in San Fernando West.

The Creole majority and its ethnic appendages in San Fernando ensured the defeat of DLP candidates in local as well as national elections. For almost two decades every member of the town council was a PNM supporter. Yet two out of the four mayors who held office in the first decade after independence were East Indian; both were prominent in the PNM, both were Christian and mixed extensively with Creoles; and both were regarded as renegades by the majority of San Fernando's East Indians.

Élite groups

Political decolonisation and the rise to prominence of non-whites in public life, coupled with the expansion of economic opportunity created by the oil industry, modified but did not radically alter San Fernando's social structure between 1920 and 1960. Segmentation of Creole from East Indian, and social divisions between Hindus, Moslems and Christians remained, but within each segment class mobility was achieved, particulary in the boom years during and after World War II. Social change was reflected by the composition of the town's élite, which became substantially multiracial as the exclusivity of whites and near-whites was relaxed.

After World War I the most prestigious element in San Fernando's society was French Creole. Largely excluded from the sugar industry by the expansion of Usine Ste Madeleine, they entrenched themselves in urban liberal professions – medicine and law. British Creole whites and expatriates – never more than about 250 families – were prominent in the professions, too, but dominated colonial administration, the business houses and executive positions in the oil industry. Gradually they admitted to their ranks – if not to their intimate society – a handful of Chinese professionals and a small number of light- and dark-brown Creoles, whose upward mobility depended on their education, administrative ability and, in some instances, their phenotype. An informant, referring to one particular family in this coloured élite, remarked: 'They are a set of people that will do anything to be a white man.' A highly acculturated East Indian group also penetrated the lower echelon of the élite. Mostly but not entirely Presbyterian (by birth or conversion), they used their primary and secondary schooling, acquired from the Canadian missionaries, to gain entry to law and medicine, previously Creole preserves.

By 1960 the French Creole élite had virtually either died out or left the town and the British élite though more numerous was greatly reduced in influence. The socially secure offspring of the coloured élite still functioned as a clique and often maintained warm personal ties with children of the Chinese and Christian East Indian élite. However, as these non-white élites became increasingly accepted as respectable, they, too, were depleted by death and emigration, especially to Port-of-Spain, though those who were left behind continued to emphasise their family background as the basis of élite status.

In this way they kept at a distance the rising Creole blacks, Hindu and Moslem traders who had amassed wealth through urbanisation and the oil boom, and rural East Indians whose land rights had turned them into oil barons. However they obtained their money, East Indians began to invest in their children's education, but in this respect the Hindus and Moslems lagged behind the Christians.

In the early 1960s, therefore, the San Fernando élite comprised the remnants of the old white 'aristocracy', non-white Creoles who had been accepted as leading families for a generation, a handful of 'old family' Chinese and a residual Christian East Indian group which was socially respectable and highly creolised. Below them ranked larger mobile strata of East Indians and blacks, who were vying for social acceptance. These blacks were professional men, politicians or civil servants – very few were businessmen. The East Indians were businessmen, professionals or political figures; one of our black professional informants commented, 'Twenty years ago they were all coolies.'

Trading activities provided the basis for non-Christian East Indian mobility in San Fernando – though it must be remembered that many East Indian professionals in the town came directly from rural, land-owning backgrounds. In 1964 there were 24 major East Indian retail outlets, all on the north side of High Street. Three, including one of the largest stores, were owned by Bombay businessmen; three belonged to Moslems; nine were owned by Hindus. Goods sold included sportswear, clothing, furniture, hardware and lumber, and jewellery; services were provided by a beautician, optician and an insurance broker. None of these business people was socially acceptable to the Creole élite, though a handful, together with some of the outstanding urban property owners, were considered to be on the fringe of that élite.

Creoles achieved a bare numerical superiority in medicine and dentistry: there were 29 Creoles in private medical practice, 27 East Indians and one *Dougla*. The Creoles were divided by colour: white expatriates (3), Creole whites (2), light coloureds (8), blacks (7), Chinese (4), Chinese–Creole (1), no information (4). Among the 27 East Indians, the Christians were the outstanding group (17), as would be expected, followed by Hindus (4), and not specified (6). There were no Moslems in private practice, but more East Indian women (4) than Creole (2) were listed as doctors. Three East Indian doctors (all Presbyterian) had white wives, but only two non-white Creoles; in none of these cases was the couple integrated into élite Creole society.

The data on dentists generally conform to the evidence for doctors. Creoles outnumbered East Indians by 11 to 4; all were men; and one non-white Creole had a white wife.

East Indians in the legal profession, however, had completely broken Creole dominance. Out of 21 barristers in 1964 only four were non-Indian – two blacks, one *Dougla* and one Chinese. Among the 17 East Indian barristers, nine were Christian (eight Presbyterian), four Hindu (including the *Dougla*), three Moslem and two unplaced. The Creole list contained no women, and only three East Indians were female. One may hypothesise that East Indian mobility had carried a handful of East Indian women on their own merits up the San Fernando social hierarchy in ways not achieved by Creole women; but successful Creole women may have chosen not to live in San Fernando. One Creole barrister was married to an East Indian wife; one Hindu was married to a white husband.

The racial balance among solicitors was more even – five Creole to eight East Indian, none of whom was a woman. The Creole solicitors were members of the old élite – one white, two Spanish–Negro (mulatto), one black and one Chinese. East Indians, once more, were predominantly Christian (five Presbyterian); two were Hindu and one was a Moslem.

Many Creoles, élite and non-élite, voiced their fears of an East Indian take-over of the town and of their corruption of public morality. A black carpenter of 60 noted, 'The young East Indians are very ambitious and are studying hard to get a good education, so as to occupy the best jobs in Trinidad, while many Negro youths don't want to study or worse to work. They prefer to spend their leisure hours idling.' A middle-aged, light-coloured businessman observed, 'Young East Indians are ruining the level and tone of the legal profession. They are people with plenty of brains but with no background and no idea of etiquette. They are all out to make money and are influenced by bribery. Trinidad is going to the dogs now the Indians are coming up.'

At the pinnacle of the social system, therefore, San Fernando moved from domination by the white élite prior to World War I to a multiracial élite in which non-white Creoles and Westernised Presbyterian East Indians occupied the lower level by World War II. At independence, attrition of the old upper stratum had produced a white or coloured Creole élite of old families and *nouveaux arrivistes* with creolised East Indian (and a few *Dougla*) appendages, and a mobile East Indian element, pressing for élite recognition, composed principally of Christians and with much smaller numbers of Hindus and Moslems (Braithwaite 1980). All these East Indian groups except the traders had a Western life-style and the formal qualifications for élite membership, and this is no doubt why Hindu and Moslem businessmen were so intent on purchasing a higher education for their children. Indeed, the movement of young East Indians into the professions – and especially into the lucrative career of barrister – reflected a calculated blend of preoccupations involving status and money-making, traditionally associated with Creoles and East Indians, respectively.

Two further points require elaboration. Why were Moslems less conspicuous in the professions than Hindus and how did this élite structure, compiled from local gazetteers and informants, compare with San Fernando entries in the 1966 version of *Who's who in Trinidad and Tobago* (Comma 1966)? Moslems were anxious to avoid proselytisation and shied away from the Canadian Mission. Hindus accepted Canadian Mission education, but, especially after 1940, rejected conversion. So Hindus moved more naturally on to the conveyor belt of primary and secondary education and professional training overseas – or, to be precise, their upper stratum did.

The 1966 *Who's who* indicated that only 6% of Trinidad's élite were resident in San Fernando. Most of the Creoles were listed because of their official positions or their rôles in prominent business houses rather than for their social standing. The inventory for East Indians was more representative, but even so included not a single Moslem. Out of 46 Creoles (72% of the total) only seven were women. Among East Indians (28% of the town's élite) only one woman was listed, and the group split into 12 Christians and seven Hindus; leading businessmen were scarcely mentioned. *Who's who* gave a view of the San Fernando élite distorted by two lenses: one the viewpoint from Port of Spain; the other, the emphasis on 'respectable' occupations rather than power or wealth. Thus East Indians were omitted because of their absence from the civil service and local government, though they commanded considerable prestige and influence, especially in their own segment of affiliation. On the theme of multiple élites, V. S. Naipaul has observed, 'each of the island's many cliques believes that it is the true élite. The expatriates believe they are the elite; so do the local whites, the businessmen, the professional men, the higher civil servants, the politicians, the sportsmen. This arrangement, whereby most don't even know they are being excluded, leaves everyone reasonably happy. And most important of all the animosity that might have been directed at the whites has been channelled off against the Indians' (Naipaul 1962, 78–9) – because they have competed with Creoles for élite positions.

The social situation in San Fernando in 1930–60 more faithfully reflected the entire Trinidadian situation than did circumstances in Port of Spain, which was almost exclusively a Creole community (Goodenough 1976, 1978). Throughout Trinidad, as in San Fernando, Christian East Indians were the outstanding leaders of the East Indian segment, and they were thrust aside only when World War II created profiteers such as Bhadase Maraj, who spearheaded the Hindu revival of the 1950s.

The rise to prominence of coloured and black élites after 1945 was a Caribbean-wide event; but San Fernando was unusual because the oil economy provided opportunities for widespread social advancement which, elsewhere, was confined to the oil-refining Dutch islands of Aruba and Curaçao. In San Fernando, however, there were two élites – the larger Creole, the smaller East Indian. Creoles were stratified by colour and family status, whereas East Indians were divided by religion. These élites were simply the top tier of a larger social pyramid whose segmented structure they reflected. It is to that larger segmented social structure and its spatial expression in the urban mosaic that attention is now directed.

4 *San Fernando: social and spatial structure in the late colonial period*

The purpose of this chapter is to explore the spatial distribution and segregation of key racial and cultural categories in San Fernando. In view of the absence of cross tabulations in the 1931 and 1960 censuses, it is important to examine the inter-relationship between such census variables as race, religion, occupation, education and family structure. Cartographic and statistical analyses of these data shed further light on the evolving social and spatial structure of San Fernando during the late colonial period, and link this urban scene to the colonial patterns of variable incorporation previously discussed.

San Fernando in 1931

The 1931 census contains no information on colour, but there is a tabulation of San Fernando's 155 expatriate whites. They were concentrated on the ridge at Spring Vale, in the residential areas looking out from Naparima Hill, and in the vicinity of Broadway and Harris Promenade (Fig. 4.1). West Indian immigrants brought in by the oil boom from the adjacent islands in the south eastern Caribbean accounted for 21.9% of the population in sections of the Spanish grid-plan and on the northern periphery, but were absent from the densely populated streets bounded on the south by Rushworth Street. Trinidadians – a catch-all term for white, brown and black Creoles – were ubiquitous, but were under-represented in the retail area, and concentrated in the high-density neighbourhoods adjacent to Cipero Street and in new suburbs facing inland across the canefields from the lower slopes of Naparima Hill. Chinese immigrants were a very small population (155 persons), all but two of whom were men. The map shows their percentage distribution by enumeration district rather than their percentage in each enumeration district: more than half were in two adjoining areas in the commercial core and the secondary shopping district around the Mucurapo street market.

West Indians and Trinidadians together made up the Creole population and accounted for 80% of the townspeople, but except for Creole whites, who probably conformed closely to the expatriate residential pattern, it is impossible to infer the distribution and spatial mix of the various colour groupings. In contrast, the data on East Indians (18% of the population) are richly detailed and in some ways superior to those available in the 1960 census (Fig. 4.2). East Indians concentrated in the southern half of the town, forming a substantial enclave bounded by Harris Promenade, Broadway, Rushworth Street and Cipero Street. Three nodes, each recording a proportion of East Indians greater

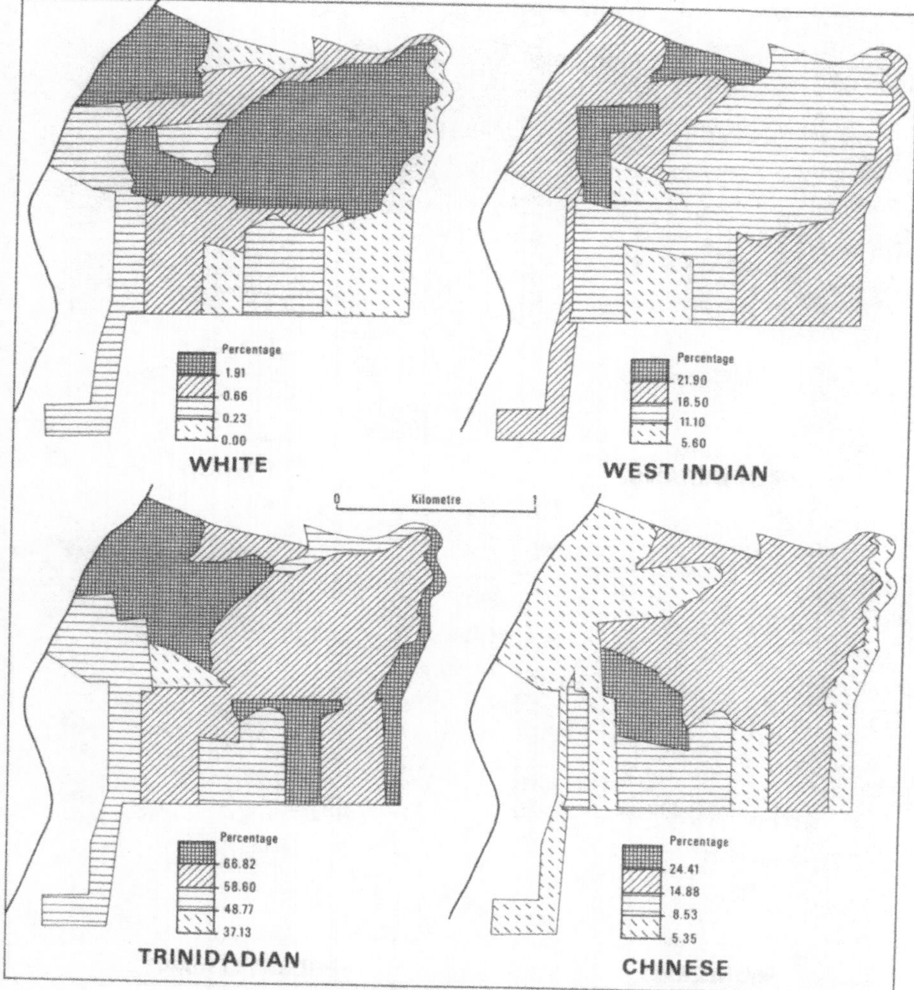

Figure 4.1 Racial categories in San Fernando, 1931.

than 25% of the enumeration district's total, were located along Broadway, south of Paradise Cemetery, and to the north of Naparima Hill along Circular Road. Three religious subsets clustered within the East Indian population: Hindus accounted for 4.4% of San Fernando's total population and concentrated in Paradise Pasture, where they carried out cattle rearing, the Spanish grid and the Cipero Street area; Moslems (5.1%) clustered around the two mosques located between Paradise Cemetery and Rushworth Street; Presbyterians (7.9%) were numerous in the streets east of Broadway and along Circular Road.

The outstanding aspect of these racial and religious distributions was the absence of ghettos. Obviously, there were major concentrations – whites clustered in salubrious neighbourhoods, East Indians dominated the southern

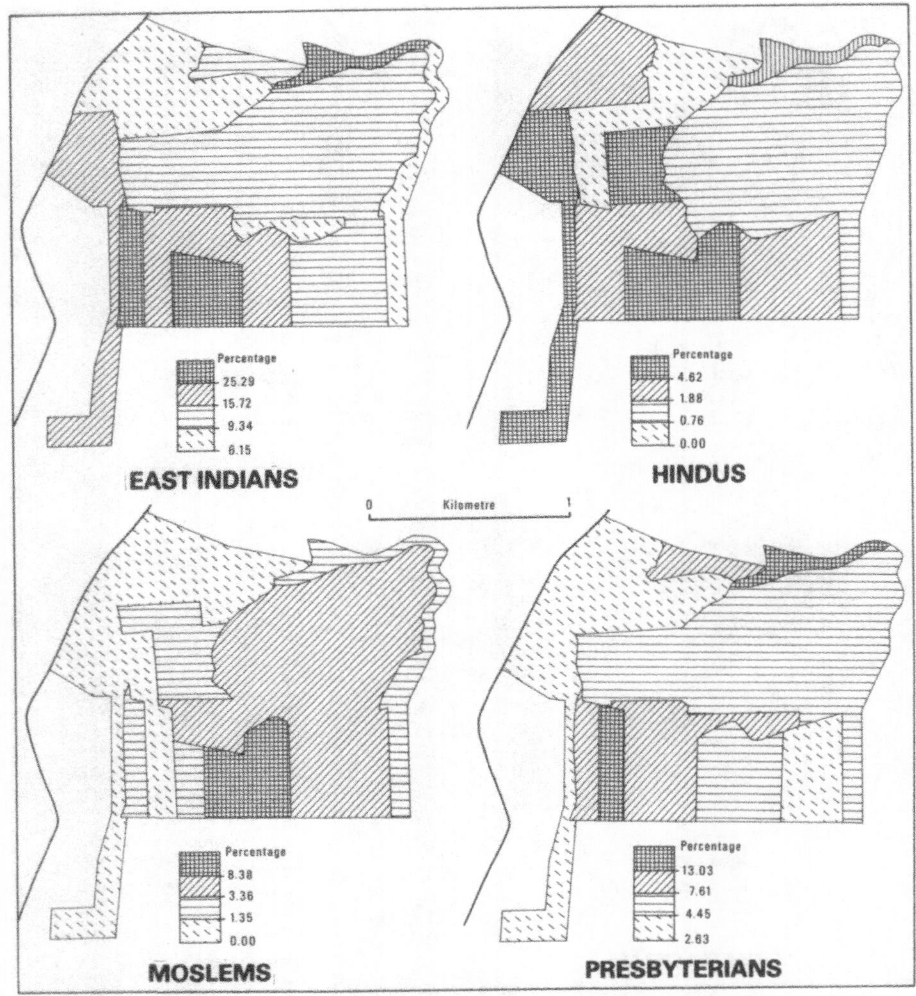

Figure 4.2 East Indian categories in San Fernando, 1931.

and western residential areas and were subdivided into small-scale religious communities – yet there was little racial exclusivity and a great deal of heterogeneity in every district. This generalisation can be tested by calculating the index of dissimilarity between groups, using enumeration district data from the census. Whereas all but one of the maps of racial and religious groups are based on the percentage of the population of each enumeration district who were, say, white or Hindu, the index incorporates the percentage distribution of each group by area. The index expresses the proportion of a particular population that would have to change their location to reproduce the percentage distribution recorded by another; it is a measure of the unevenness of distribution among specified categories.

The indices of dissimilarity are socially revealing in three respects (Table 4.1). Expatriate whites were highly segregated, and more 'distant' from Hindus (71.7) and Moslems (76.5) than from the rest; Chinese immigrants were highly segregated from all others, except Presbyterian East Indians; lower indices of dissimilarity were recorded among the East Indian religious populations, especially Hindus and Presbyterians (35.4), though the indices confirm that there clearly was moderate segregation between Moslems and Hindus on the one hand (43.7) and Moslems and Presbyterians (44.6) on the other. These tendencies towards segregation reflected white élitism, Chinese commercialism, and subtle relationships of social proximity and distance among the East Indians, structured around Moslem isolationism and the close family bonds linking Hindus and Christian East Indians. When the ill-defined Trinidadian and West Indian groups are combined to form a Creole segment, their ubiquity is exposed in low to moderate indices of dissimilarity from Hindus (34.8) and Moslems (43.5), though they were more markedly segregated from Chinese (49.8) and expatriate whites (53.9). Presbyterians already formed an intermediate group: they were closer to coloured Creoles (21.7) than to Hindus (35.4) and Moslems (44.6), and recorded a low index of dissimilarity, as compared with others, with expatriate whites (55.5).

Despite reservations about the utility of spatial analysis, given imprecise census categories and the paucity of census units (Appendix A), a clear and consistent pattern of racial and religious segregation emerges from the 1931 data. To understand more fully the basis of these residential patterns it is essential also to examine census data for housing, Christianity and various other cultural indicators. This done, correlations among race, religion, housing and marriage can be set forth.

The information about housing is far from complete: it excludes details about dwelling size and type, but, nevertheless, certain cautious generalisations can be made (Fig. 4.3). Divided houses – tenements – were commonest (more than 28% of the accommodation) in the commercial centre and in the Moslem section of the town. Barracks or ranges were heavily concentrated in the western neighbourhoods and coincided with the distribution of Hindus, espe-

Table 4.1 Indices of dissimilarity for racial and religious groups in San Fernando, 1931 (*source:* Census of Trinidad & Tobago, 1931).

Group	Expatriate white	East Indian	Hindu	Moslem	Presbyterian	Chinese	West Indian and Trinidadian
expatriate white	—	64.4	71.7	76.5	55.5	58.5	53.9
East Indian		—	19.3	33.1	22.3	56.6	30.7
Hindu			—	43.7	35.4	55.5	34.8
Moslem				—	44.6	58.9	43.5
Presbyterian					—	48.3	21.7
Chinese						—	49.8
West Indian and Trinidadian							—

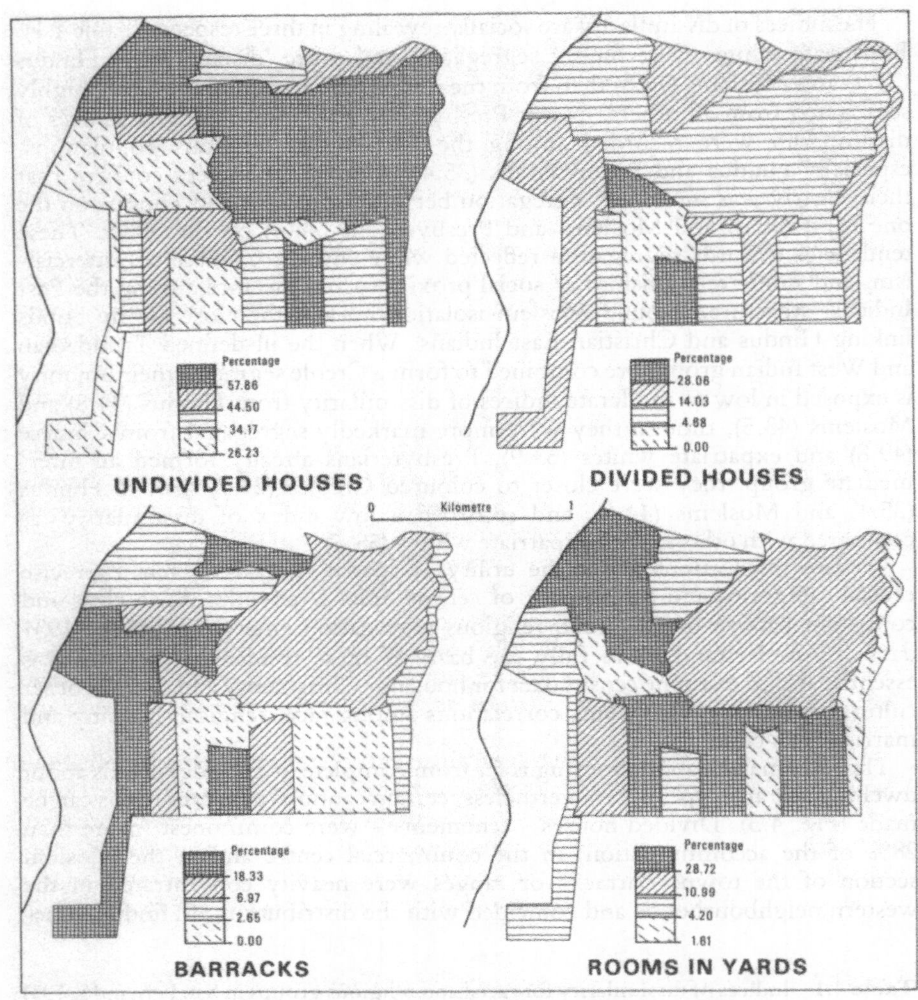

Figure 4.3 Housing types in San Fernando, 1931.

cially in Paradise Pasture. Only here, however, did they account for as much as 18% of the accommodation, for barracks were a rural type of dwelling and associated with the period of indentureship. By contrast, rooms in yards were essentially an urban housing type, comprising small huts constructed on rented spots, scattered throughout the town, and complementing the tenements in the densely populated southern neighbourhoods between Coffee Street and Rushworth Street. The term 'undivided houses' covered a multitude of types, ranging from merchants' mansions to small wooden huts. Undivided houses were most widespread in the north and east of San Fernando (more than 57.9% of dwellings), but were common also in the southern residential area, notably in districts where Presbyterians concentrated.

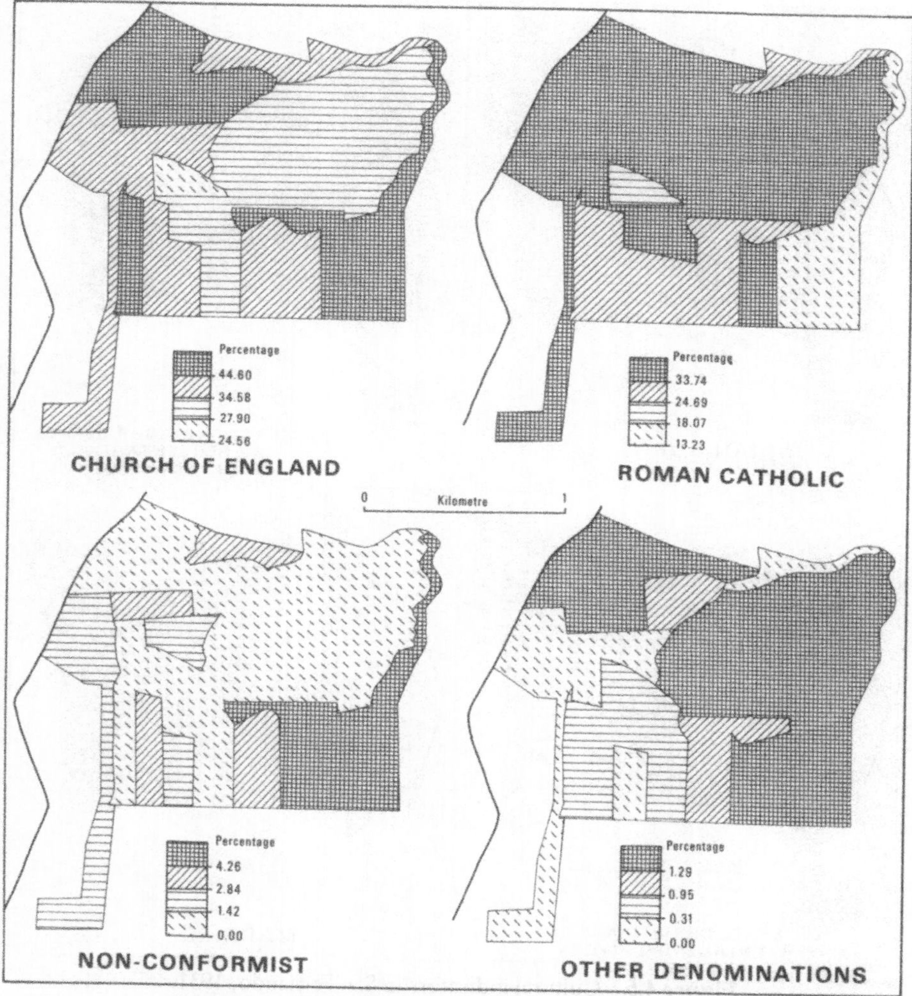

Figure 4.4 Christian religions in San Fernando, 1931.

Evidence relating to the 19th century has already underscored the social significance of religious affiliation in Creole San Fernando (Fig. 4.4). Members of the Church of England accounted for 41.4% of the town's population, most of them located in the lower status eastern neighbourhood; few lived in the commercial centre, though the Spring Vale ridge was occupied by a small cluster. Roman Catholics, 30.6% of San Fernando's population, lived on the Spring Vale ridge, but concentrated where Anglicans were scarce. Non-Conformists repeated the Anglican distribution, but with smaller percentages, and 'other denominations' – probably Afro-Christian groups – formed a tiny majority in the north and east, spatially distant from East Indians in the town's ecology.

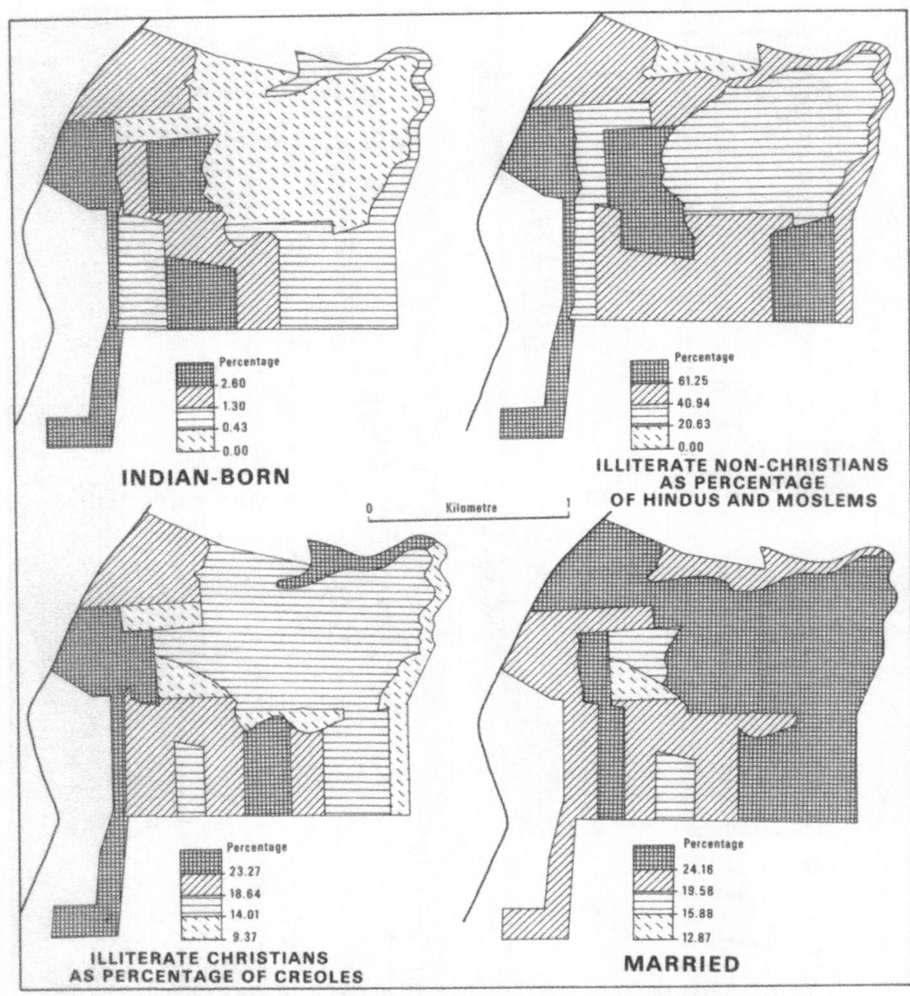

Figure 4.5 Cultural indicators in San Fernando, 1931.

Three additional cultural features were important in the social fabric of San Fernando (Fig. 4.5). At the time of the 1931 census, indentured immigration had been abolished for only 14 years, and there were still 262 Indian-born residents. They accounted for more than 2.6% of the population only in Paradise Pasture, in the commercial centre, where many Bombay businessmen resided, and north of Rushworth Street. In short, they concentrated, but at low levels of intensity, in the East Indian sections in the south and west of the town.

Wherever East Indians were prevalent, illiteracy in English among non-Christians was over 60%. Illiteracy among Christians, however, was more prevalent along Circular Road, Cipero Street and in Paradise Pasture than in Spring Vale and the commercial thoroughfares of High Street and Coffee Street.

Table 4.2 Matrix of Spearman rank correlation coefficients for selected variables in San Fernando, 1931 (*source:* Census of Trinidad & Tobago, 1931; —— highest coefficient in each column, ----- second highest coefficient in each column).

Group	White	Hindu	Presbyterian	Moslem	Rooms in yards	*Douglas*	Church of England	Roman Catholic	Married	China-born	India-born	Trinidad-born East Indian	Barracks
white	—	−0.21	−0.22	−0.14	0.05	−0.09	0.03	0.25	0.30	0.35	−0.04	0.02	0.24
Hindu	−0.21	—	0.35	−0.61	0.21	−0.33	−0.38	−0.43	−0.49	0.15	0.71	0.67	0.07
Presbyterian	−0.22	0.35	—	0.14	0.33	0.10	−0.20	−0.16	0.02	0.20	−0.07	0.51	0.03
Moslem	−0.14	−0.61	0.14	—	0.26	−0.35	−0.37	−0.15	−0.21	0.34	0.34	0.37	0.39
rooms in yards	0.05	0.21	0.33	0.26	—	−0.01	0.14	0.01	0.29	0.06	−0.04	0.15	−0.53
Douglas	−0.09	−0.33	0.10	−0.35	−0.01	—	0.12	0.85	0.22	0.23	0.30	0.87	0.67
Church of England	0.03	−0.38	−0.20	−0.37	0.14	0.12	—	−0.17	0.41	−0.63	−0.49	−0.35	−0.26
Roman Catholic	0.25	−0.43	−0.16	−0.15	0.01	0.85	−0.17	—	0.07	0.05	−0.06	0.35	0.17
married	0.30	−0.49	0.02	−0.21	0.29	0.22	0.41	0.07	—	−0.38	−0.55	−0.16	0.32
China-born	0.35	0.15	0.20	0.34	0.06	0.23	−0.63	0.05	−0.38	—	0.22	0.07	−0.14
India-born	−0.04	0.71	−0.07	0.34	−0.04	0.30	−0.49	−0.06	−0.55	0.22	—	0.39	0.16
Trinidad-born East Indian	0.02	0.67	0.51	0.37	0.15	0.87	−0.35	0.35	−0.16	0.07	0.39	—	0.89
barracks	0.24	0.07	0.03	0.39	−0.53	0.67	−0.26	0.17	0.32	−0.14	0.16	0.89	—

All East Indian districts were markedly isolated from neighbourhoods where marriage was the norm.

To trace these inter-relationships more systematically, 13 census variables reflecting various characteristics of East Indians (5), racial minorities (3), religion (2), housing (2) and marriage (1) have been selected and intercorrelated thus producing a matrix of ecological inter-relationships for San Fernando (Table 4.2). Many of the variables are either skewed or bimodal, and a non-parametric technique for measuring association (Spearman rank correlation) has been used. By inspecting the correlation matrix it is possible to examine the relationships between whites, for example, and other variables which have been mapped. However, the object in the first instance is to establish the broad pattern, and this can best be achieved by linkage analysis.

The highest and second highest coefficients, irrespective of sign, in each column of the correlation matrix have been underlined, and a linkage diagram based on these correlations has been constructed to show the nature, strength and direction of these major bonds (Fig. 4.6). The groups created by linking the highest coefficients are characterised by the fact that each variable in any group is more highly correlated with another variable in the same group than it is with any variable in any other (Fig. 4.6). In almost all cases the second highest correlation for any variable also falls in the same group, and, in some instances, variables which are members of the same group are themselves highly interlinked, with little or no significant correlation outside.

The linkage diagram depicts one large constellation binding together Trinidad-born East Indians, Hindus and barracks; negatively correlated with it is a smaller offshoot based on married people (Fig. 4.6). The former constellation picks out major features of the East Indian population, and the latter reflects aspects of high social status. Using the highest and second highest coefficients for each variable, four nodes may be identified. In rank order of the number of correlation bonds involved, these nodes are Hindus, Trinidad-born East Indians, barracks and married people. Examination of these nodes and their satellite variables provides a preliminary survey of San Fernando's social structure in 1931.

Hindus were positively associated with the India-born (0.71) and Presbyterians (0.35), and negatively correlated with Moslems (−0.61), married people (−0.49) and Roman Catholics (−0.43). Trinidad-born East Indians recorded many positive links – with barracks (0.89), *Douglas* (0.87), Hindus (0.67) and Presbyterians (0.51). Barracks were positively correlated with Trinidad-born East Indians (0.89), *Douglas* (0.67) and Moslems (0.39) and negatively with rooms in yards (−0.53). Care must be taken in interpreting these results, since Moslems and Hindus were subsets of the East Indian population. Of lesser statistical importance but of great social interest was the married category, which was linked negatively to the India-born (−0.55) but positively to the whites (0.30). Attention must also be drawn to the in-between nature of the *Douglas*, who were linked by Catholicism (0.85) to the Creoles and by barrack residence (0.67) to the East Indians: almost 60% of the *Douglas* had East Indian fathers.

This analysis brings the East Indians into sharp focus. As for the Creole majority, three variables can be treated as guides to likely Creole characteristics

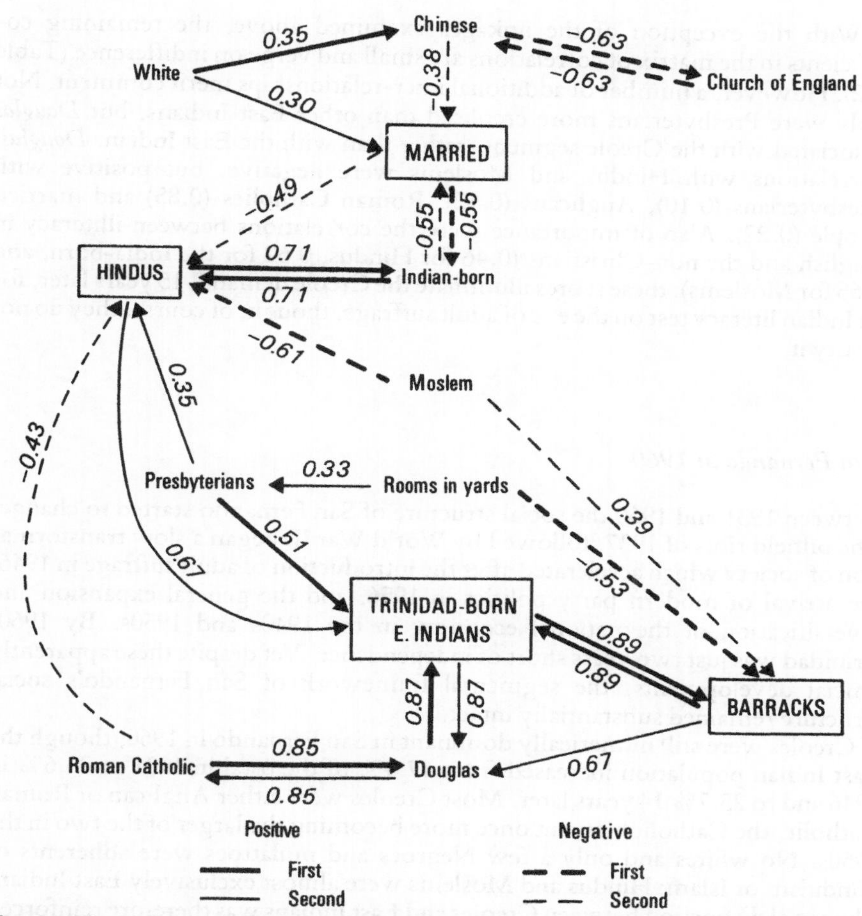

Figure 4.6 Linkage of highest and second highest correlation for each variable in San Fernando, 1931.

– Anglicanism, Roman Catholicism and marriage – but all denote high social status and are blunt instruments for separating blacks from browns. As we have already seen, official marriage was a key factor in distinguishing between whites and East Indians, as traditional Hindu and Moslem weddings lacked official recognition. Moreover, Catholicism was interestingly linked to *Douglas* (0.85) and Trinidad-born East Indians (0.35) and negatively correlated with Hindus (−0.43). Anglicanism, on the other hand, was not a discriminant variable of central significance, though it did have a modest correlation (0.41) with marriage and hence with the Creole stratification. However, when all three 'Creole' characteristics – Anglicanism, Roman Catholicism and marriage – are correlated with the principal East Indian variables – Hindu, Presbyterian, Moslem and India-born, 11 out of the 12 coefficients are negative, and this is a clear expression of Creole–East Indian pluralism.

With the exception of the linkages examined above, the remaining coefficients in the matrix of correlations are small and verge on indifference (Table 4.2). However, a number of additional inter-relationships merit comment. Not only were Presbyterians more creolised than other East Indians, but *Douglas* associated with the Creole segment, rather than with the East Indian. *Douglas*' correlations with Hindus and Moslems were negative, but positive with Presbyterians (0.10), Anglicans (0.12), Roman Catholics (0.85) and married people (0.22). Also of importance were the correlations between illiteracy in English and the non-Christians (0.46 for Hindus, 0.62 for the India-born, and 0.65 for Moslems): these scores illuminate the Creole demand, 15 years later, for an Indian literacy test on the eve of adult suffrage, though, of course, they do not justify it.

San Fernando in 1960

Between 1931 and 1960 the social structure of San Fernando started to change. The oilfield riots of 1937, followed by World War II, began a slow transformation of society which accelerated after the introduction of adult suffrage in 1946, the arrival of modern party politics in 1956, and the general expansion and diversification of the national economy in the 1940s and 1950s. By 1960, Trinidad was just two years short of independence. Yet despite these apparently crucial developments, the segmental framework of San Fernando's social structure remained substantially intact.

Creoles were still numerically dominant in San Fernando in 1960, though the East Indian population increased from 17.4% of the total in 1931 to 23.6% in 1946 and to 25.7% 14 years later. Most Creoles were either Anglican or Roman Catholic, the Catholic element once more becoming the larger of the two in the 1950s. No whites and only a few Negroes and mulattoes were adherents of Hinduism or Islam: Hindus and Moslems were almost exclusively East Indian. The racial distinction between Creoles and East Indians was therefore reinforced by religious differences. Moreover, most East Indian converts were Presbyterian rather than Catholic or Anglican; by 1960 nearly all Presbyterians were East Indian, apart from 340 persons of mixed race who were presumably *Douglas* (Table 4.3).

Caribs, Syrians, Portuguese and Chinese traditionally stood outside the Creole colour-class hierarchy. The Syrians and Portuguese were Catholic, and the Chinese community comprised a Catholic majority and an Anglican minority. In religion, then, these groups were aligned with the Creoles. Furthermore, through their success in trade, the Syrians, Portuguese and Chinese gradually gained access to the upper reaches of the Creole stratification.

Maps have been prepared to show the distribution of the major racial and religious groups, using as a basis the 47 enumeration districts into which the town was divided for the 1960 census. Portuguese, Syrians, Caribs and *Douglas* do not appear in the enumeration district tabulations and cannot be considered in the analysis which follows.

Whites comprised just over 3% of San Fernando's population in 1960, almost

Table 4.3 Race and religion in San Fernando, 1960 (*source:* Census of Trinidad & Tobago, 1960).

Race	Total and % by race	Religion (no.; as % of racial group in parentheses) Anglican	Methodist	Presbyterian	Roman Catholic	Hindu	Moslem	Other*
Negro	18 784 (46.9)	8639 (45.8)	878 (4.7)	37 (0.2)	7199 (38.2)	6 (0.03)	11 (0.06)	2014 (10.7)
white	1306 (3.3)	422 (32.5)	21 (1.6)	64 (4.9)	649 (50.0)	0 (0.0)	0 (0.0)	150 (11.6)
Portuguese	140 (0.4)	1 (0.7)	0 (0.0)	7 (5.0)	131 (93.0)	0 (0.0)	0 (0.0)	1 (0.7)
East Indian	10 296 (25.7)	366 (3.6)	10 (0.1)	3379 (32.8)	848 (8.2)	3011 (29.2)	2282 (22.1)	400 (3.9)
Chinese	705 (1.8)	166 (23.2)	0 (0.0)	11 (1.5)	473 (66.2)	1 (0.1)	0 (0.0)	54 (7.6)
mixed	8283 (20.7)	1949 (23.5)	134 (1.6)	344 (4.2)	5400 (65.1)	26 (0.3)	24 (0.3)	406 (4.9)
Carib	12 (0.0)	6 (50.0)	0 (0.0)	1 (8.3)	5 (41.5)	0 (0.0)	0 (0.0)	0 (0.0)
Syrians	152 (0.4)	10 (6.6)	0 (0.0)	0 (0.0)	135 (88.8)	0 (0.0)	0 (0.0)	7 (4.6)
other	146 (0.4)	17 (11.5)	0 (0.0)	0 (0.0)	109 (74.1)	0 (0.0)	1 (0.0)	19 (12.9)
not stated	6 (0.0)	3 (50.0)	0 (0.0)	0 (0.0)	0 (0.0)	3 (50.0)	0 (0.0)	0 (0.0)
Total	39 830 (100.0)	11 579	1043	3843	14 949	3047	2318	3051

*The 'other religion' category consists of Christian religious groups not listed separately in the table, of other non-Christians, and of those not stated or with no religion.

the same proportion as in 1946 (Table 3.1), and were heavily concentrated at St Joseph's Village and in the old white neighbourhood of Spring Vale (Fig. 4.7). Other whites lived in postwar suburbs on the northern and southern peripheries of the town and in older housing flanking Naparima Hill. Whites were virtually absent from an east–west belt that ran through the centre of the town comprising the central business district, various governmental and educational institutions, and government-financed housing schemes.

Negroes accounted for 47% of the inhabitants in 1960, fewer than the 54.7% recorded in 1946, and were heavily concentrated in postwar government housing on the western edge of the town. The white enclaves and the neighbourhoods surrounding the temples and mosques acted as poles of repulsion for the blacks.

The mixed group or coloureds comprised 21% of the population – a substantial increase over the 16% in 1946. Its members occupied an intermediate position, socially and geographically, between the black and white populations. Some lived in the government housing schemes, others in élite Spring Vale, but most were located on the northern and western flanks of Naparima Hill.

The Chinese, local and foreign-born, accounted for just under 2% of San

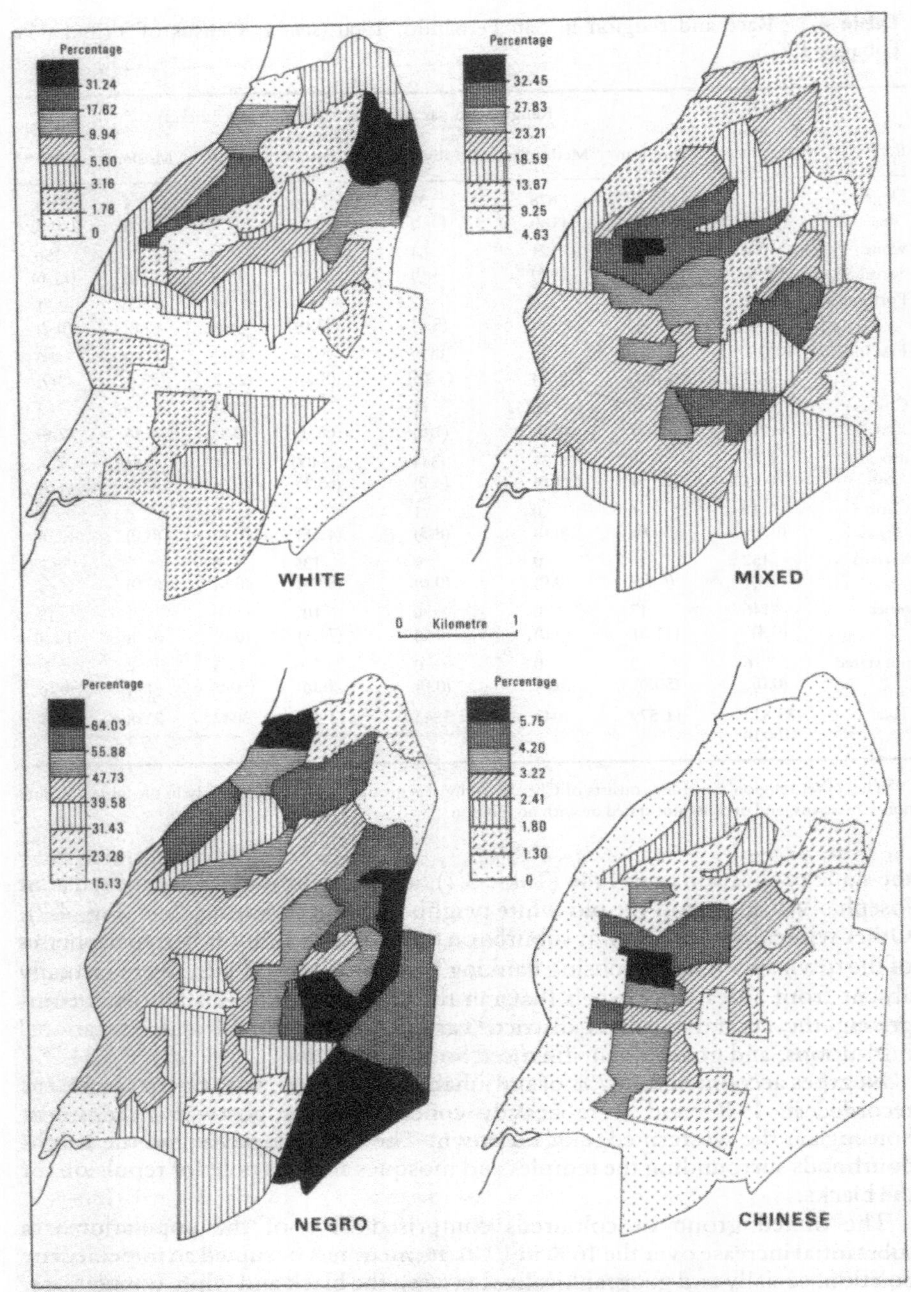

Figure 4.7 Racial distributions in San Fernando, 1960.

Fernando's inhabitants in 1960. They were strongly associated with the commercial centre of the town – as in 1931 – and concentrated on High Street, Coffee Street, Pointe-à-Pierre Road and, above all, on Mucurapo Street. This pattern differed from the residential distribution of all other groups, because the majority of Chinese families were retailers and many lived over their shops. However, there was some evidence of Chinese penetration of the inner suburbs, including Spring Vale, but again largely as shopkeepers.

East Indians comprised 26% of the town's inhabitants. They resided in western districts, as in 1931, especially in new residential areas on the northern and southern peripheries (Fig. 4.8). Wherever Negroes were numerous, East Indians failed to account for a quarter of the inhabitants.

Hindus (8%) were not the largest element in the East Indian population, but their distribution most faithfully reflected the pattern of the group as a whole, and their absolute and proportional contributions to the town's population increased substantially between 1946 (1788) and 1961 (3047). Hindus clustered on the edge of the town, especially in the newest suburbs. Their highest concentration was at Kakatwey, the shanty town, but some also occupied part of the high-ranking neighbourhood at St Joseph's Village. Hindus were well represented in several of the government housing schemes, but were conspicuously absent from some major Creole districts in the north eastern section of San Fernando.

Moslems (6%) likewise concentrated on the northern and southern boundaries of town, but few lived in the shanty town. They congregated in the district between Coffee Street and Rushworth Street which was so prominent an Indian area in 1931. Unlike Hindus, their proportional contribution to the town's population declined slightly between 1946 and 1960, though their numbers increased from 1952 to 2318.

The proportional contribution of Presbyterians to San Fernando's population remained stable between 1946 and 1960 (10%) and by the latter date they were strongly associated with the northern side of Naparima Hill, which they had colonised by 1931, and with the area to the west of Broadway, where the Canadian mission had established educational institutions. Presbyterians were notably absent from the shanty town, from the élite areas, from government housing schemes, and from most districts with heavy Hindu, Moslem and Negro populations.

For each racial and religious category, as in 1931, there were nodes in which the category in question was over-represented compared with its proportional contribution to the total population of the town. The nodes were surrounded by areas where each category was under-represented. Yet except for the large Negro population few of these nodes were predominantly of any single category. It is easy enough to describe the government housing schemes at Navet and Pleasantville as Negro areas; to think of St Joseph's Village and Spring Vale as white enclaves; to locate many of San Fernando's 'old' coloured families, whose fretwork verandas overlooked the Gulf of Paria from the lower slopes of Naparima Hill; or to point to the concentration of East Indians on the northern and southern peripheries of the town. However, the fact remains that in only four enumeration districts, 8% of the total, did the white, mixed or East Indian populations, taken separately, comprise more than half the inhabitants.

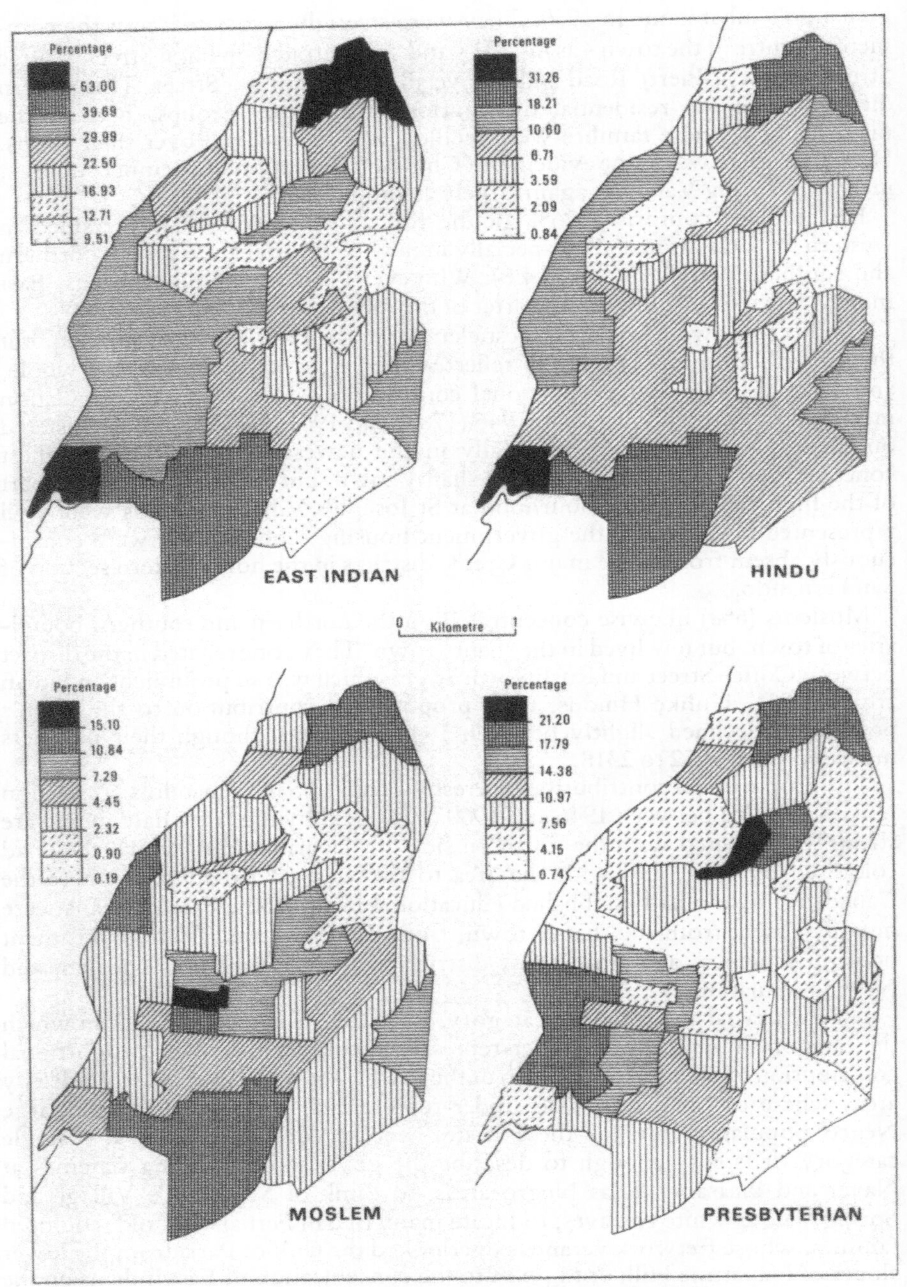

Figure 4.8 East Indians in San Fernando, 1960.

Even in the enumeration district that was most Chinese in population, 19 out of 20 persons were either Creole or East Indian.

The absolute size of each group was, of course, an important factor. Negroes, who accounted for almost half the population of San Fernando, were the only racial element numerically to dominate several adjacent districts. Yet even they exceeded 70% in only two census areas. Hindus formed a majority of the population in only one enumeration district, and Moslems, Presbyterians, whites and Chinese in none. Nevertheless, the areas of relative concentration were important for the religious organisation of the East Indians: Hindu nodes, at the northern and southern ends of the town, supported temples, and the major Moslem districts had two mosques.

The absence of minority group concentration had an important corollary. Except for the small population of whites, who were absent from one-quarter of the districts, the remaining groups were represented in each census area. A major consequence of these distributions was that although Creoles dominated all but two enumeration districts, the locations of Creoles and East Indians were polarised rather than segregated; this pattern, too, had existed in 1931.

This generalisation about the lack of segregation can be examined by means of indices of dissimilarity. The index of dissimilarity between Creoles and East Indians was 27.0, which means that 27.0% of the East Indian population would have had to change their residence to reproduce the percentage distribution recorded by the Creoles and vice versa (Table 4.4). Although the index relates only to the proportion and not to the absolute size of groups, it nevertheless provides a useful summary index. Moreover, in this instance it substantiates the conclusion drawn from the maps – namely that Creoles and East Indians are only weakly segregated (Clarke 1971).

It may be argued that this conclusion depends upon a specific areal base, the enumeration district, and that a change in the scale of the analysis might yield different results. To test this possibility the electoral rolls were searched for the addresses of East Indians, and voters' photographs in the electoral register examined to check East Indian identification by name. The register showed that East Indians could be selected accurately by name, while the rolls indicated that

Table 4.4 Indices of dissimilarity for racial and religious groups in San Fernando, 1960 (*source:* Census of Trinidad & Tobago, 1960).

Group	Mixed	Negro	Hindu	Moslem	Presbyterian	East Indian	Chinese
Creole			33.7	34.2	27.9	27.0	
white	59.5	57.6	61.0	63.8	59.0	59.2	70.2
mixed		21.6	31.5	30.5	24.2	22.4	36.0
Negro			32.2	34.5	27.4	27.2	41.9
Hindu				30.3	28.0		49.4
Moslem					30.5		37.4
Chinese					37.3		

even where East Indians comprised more than 25% of the population, there were no streets devoid of Creoles.

Although degrees of spatial association between racial and religious groups may be inferred from the maps, these, too, are easier to assess from indices of dissimilarity. The whites remained highly segregated from all the other groups but were closer to the Negroes (57.6), the East Indians (59.2) and the mixed (59.5) than to the Chinese (70.2). Among the East Indians, the distribution of whites was more nearly approximated by the Presbyterians (59.0) than by the Hindus (61.0) and the Moslems (63.8). These religious groups were much more segregated from the whites than they were from one another.

The high index of dissimilarity between whites and Hindus is expressed cartographically by the enclaves of whites at St Joseph's Village and Spring Vale and by the concentration of Hindus in the squatter camp at Kakatwey. By contrast, all the non-white groups except the Chinese were inter-related by indices of dissimilarity ranging from 21.6 to 34.5, or almost half those recorded for whites. This shows the continuing isolation of the white population and hints at its social supremacy in 1960.

It is clear that the Negro and mixed populations were closer to one another in distribution than to either the entire East Indian population or any one of the East Indian groups. In every instance the mixed population (coloured Creoles) was closer than Negroes to the Hindus, Moslems and Presbyterians, though both these Creole categories recorded low indices of dissimilarity with the Presbyterians. Although Negroes were closer in distribution to Hindus than to Moslems, the mixed group was closer to Moslems than to Hindus. When it is remembered that together the Negro and mixed populations accounted for about 95% of the Creole inhabitants of San Fernando, it is hardly surprising that the index of dissimilarity for Creoles of all colours increased from 27.9 with Presbyterians to 33.7 and 34.2 with Hindus and Moslems respectively. In no case were the indices large.

So far as the East Indian groups were concerned, the indices of dissimilarity for Hindus and Moslems, taken separately, were greater with the Creoles than with any one East Indian element (Table 4.4). Furthermore, Moslems were marginally closer to Hindus (30.3) than to Presbyterians (30.5), whereas Hindus were closer to Presbyterians (28.0) than to Moslems (30.3), a pattern which is significant largely because it is identical to that recorded in 1931. The anomalous position of the Presbyterian East Indians is emphasised by the small indices of dissimilarity with the mixed group (24.2) and with Negroes (27.4), figures even below their indices with Hindus (28.0) and Moslems (30.5). However, Moslems were no more dissimilar from Hindus (30.3) than from Presbyterians (30.5).

The socio-economic basis for these spatial patterns and associations deserves brief attention. The 1960 census of Trinidad published no cross tabulation between race and occupation for San Fernando. To remedy this, indices of dissimilarity were calculated between males in non-manual occupations and the total membership of the key racial and religious categories in San Fernando. The higher the index of dissimilarity the greater the divergence of the relevant category from the spatial pattern of white-collar workers, and the lower its socio-economic status.

The two major segments, Creoles (19.2) and East Indians (17.5), achieved similar indices, though Creoles recorded a slightly lower socio-economic status. Within the Creole population the mixed group (14.0) ranked above Negroes (21.5). It is pointless to compute the index for whites because they were so highly segregated. However, the map shows that the principal white enclaves recorded the highest proportions of males in non-manual occupations (Fig. 4.9). Among East Indians, Presbyterians (20.0) ranked above Hindus (27.5) and Moslems (27.6).

These data depict a Creole population among whom whites, coloureds and Negroes ranked in decreasing order of socio-economic status and increasing order of numerical size, and an East Indian population of parallel status that comprised a relatively high-ranking stratum of Presbyterians and low-ranking strata of Hindus and Moslems. Although Hindus and Moslems in general occupied the lowest socio-economic positions in San Fernando, the Presbyterians *en bloc* ranked socio-economically above the Negroes. The success of Presbyterians in obtaining white-collar jobs highlights the importance of conversion as a factor in East Indian social mobility, and it was largely through the achievements of members of this particular religion that East Indians ranked marginally above Creoles. Racial and religious categories adjacent to one another in the socio-economic scale in most cases tended to live close to one another spatially. Similar socio-economic scales imply similar jobs and in this instance equal opportunities to obtain housing.

Yet few of the spatial patterns are explicable entirely in economic terms, nor is it always clear whether the patterns of association are causal or coincidental. Cultural factors undoubtedly play a part; for example, the greater spatial proximity of Hindu with Presbyterian than of Hindu with Moslem may be attributable to greater antipathy between members of the two Eastern religious groups and above all to the fact that most of the Christian converts had been Hindu. Many Presbyterians retained cultural practices of Hindu origin, along with family and personal links with Hindus.

People of mixed descent lived closer to Hindus, Moslems and Presbyterians than did Negroes. This is largely explained by the high proportion of Negroes who lived in government housing schemes; all other groups were virtually confined to the private sector. It is harder to explain why Negroes more closely reproduced the spatial distribution of Hindus than Moslems or why the mixed group was closer to Moslems than to Hindus. It is important to remember that the East Indian religious groups are to a marked degree internally organised. The presence of religious institutions in San Fernando both influenced and was influenced by the location of considerable numbers of Hindus, Moslems and Presbyterians. These groups were essentially inward-looking, and their relationships with other groups must have been, to some extent, coincidental.

A further factor should also be borne in mind: the East Indian suburbs on the northern and southern extremities of the town were less than five years old in 1960. They emerged from the rapid increases in the East Indian population of San Fernando and from the growing demand for homes in 'Indian' areas developed by East Indian realtors. The resulting pattern of suburban growth expressed the increasing tension between East Indians and Creoles, which in turn reflected the deep involvement of race in Trinidad's politics. These newer suburbs thus contained large numbers of socially mobile Hindus, Moslems and

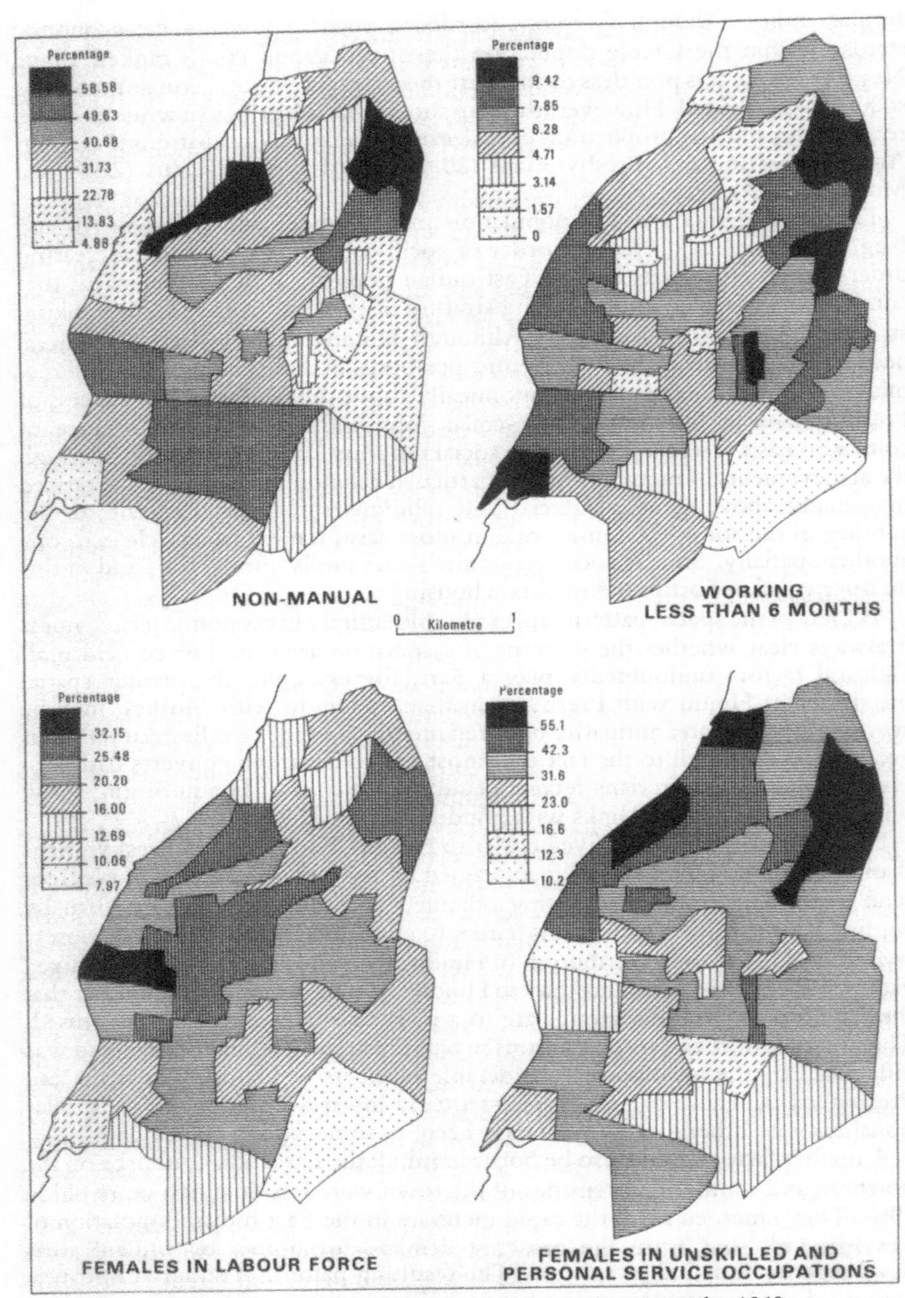

Figure 4.9 Class and the labour force in San Fernando, 1960.

Presbyterians, and their growth also influenced the indices of dissimilarity among the East Indians. As they developed at the same time as the government housing schemes, they helped to sustain segregation between Creoles and East Indians.

Spatial analysis shows considerable racial and religious intermixing, underlain by socio-economic factors. From this conclusion a continuum can be developed that provides a summary of the socio-spatial situation in San Fernando in 1960 and, by extrapolation, in 1931 as well. Whites constituted a segregated minority located at the apex of the social scale distanced spatially and socio-economically from all the other categories – a feature typical of colonial towns (King 1976). The Negro and mixed populations who comprised the greater part of Creole society, were closer to one another than to any other category. Hindus and Moslems formed distinct elements, generally ranking at the bottom of the social scale. Between them and the Creoles, but closer to the Creoles, were the Presbyterian East Indians. Though by no means completely acculturated to the Creole community, they stood closer – culturally if not geographically – to the Hindus than to the Moslems in this continuum.

The spatial data suggest the absence of marked segregation except for whites; the new East Indian suburbs and the concentration of Negroes in the government housing schemes were the closest approximation to coloured ghettos. It is likely that spatial separation and segregation are associated with size; unlike large cities, small, compact towns such as San Fernando tend toward low indices of dissimilarity. Furthermore, the history of the town scarcely encourages segregation: representatives of all the groups have been residents in San Fernando for at least half a century and in most cases for more than a century. Nonetheless, social differentation *was* reflected in the urban mosaic. Based on differences in the labour force, family, religion and education, social areas can be depicted.

The distribution of males of working age in professional and managerial occupations reflects the high status of St Joseph's Village and Spring Vale, the low rank of Kakatwey and some of the government housing schemes, and the middle-class character of Naparima Hill and the suburbs on the northern, western and southern peripheries of the town (Fig. 4.9). A mirror image of this pattern emerges in the map of long-term unemployment, though St Joseph's Village accommodated some of the wealthiest and the most indigent townspeople.

More than one-fifth of the entire female population of San Fernando was employed in 1960. This fraction rose to over one-quarter in the commercial area, and to one-third in neighbourhoods adjoining the hospital. St Joseph's Village and Spring Vale also recorded high rates of female economic activity, with large numbers of living-in maids. Moreover, high proportions of women in domestic service typified most districts of median status.

Officially recognised marital unions, everywhere more common than concubinage, were especially associated with areas of high social status, like St Joseph's Village and the adjacent districts, and comparatively rare in the town centre (Fig. 4.10). Rushworth Street formed a major behavioural divide: to the south lawful unions were much more common than to the north.

Consensual or common-law unions were ubiquitous but not the norm – a

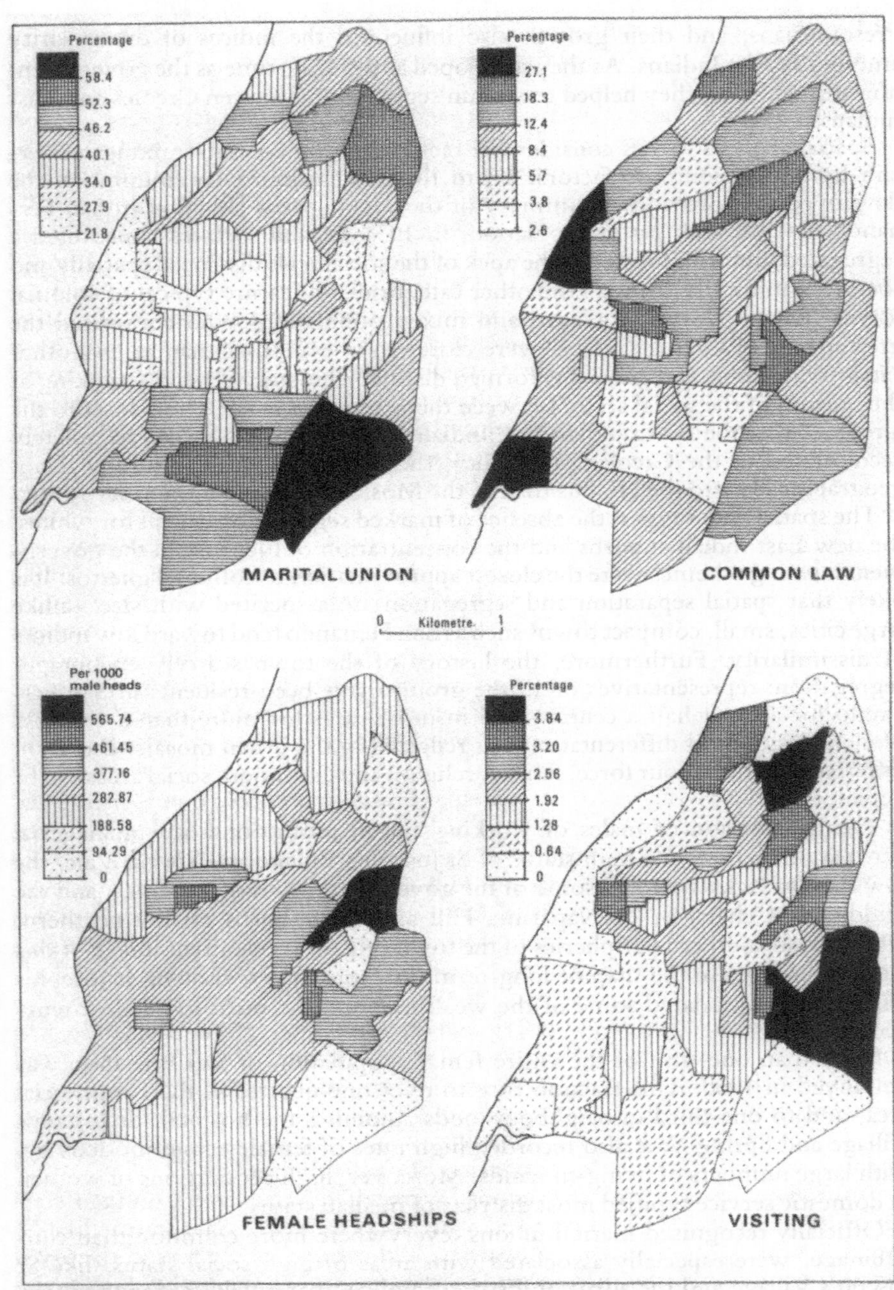

Figure 4.10 Family structure in San Fernando, 1960 (Data for women only).

surprising feature for a Caribbean town. Their spatial pattern is difficult to interpret, but common-law unions were frequent at Kakatwey, in the town centre, on the edge of Naparima Hill, and in some of the government housing projects.

Female headship characterised at least one-fifth of the households in most areas of San Fernando. Slightly higher rates occurred in some of the lowest ranking neighbourhoods, though both St Joseph's Village and Kakatwey, which formed the apex and base of the social pyramid respectively, recorded low proportions of female heads. Female headship was rarely associated with marital or consensual unions, but was strongly linked with women who were engaged in visiting relationships, though only small numbers were involved in this system of mating.

Hinduism and Islam were racially exclusive of blacks and institutionally organised, and their East Indian adherents possessed a strong sense of cultural identity, despite weak segregation (Fig. 4.8). Christianity was less of an exclusive social bond, though there were significant residential distinctions between Protestants and Catholics, who between them accounted for 80% of the town's population (Fig. 4.11). Protestants dominated the poorer areas on the eastern periphery, though most middle-class neighbourhoods recorded substantial numbers. Catholics were numerous around Naparima Hill, where they had clustered in 1931, and at Spring Vale and St Joseph's Village, but also concentrated in the government housing schemes. At the opposite end of the spectrum of religious respectability from the denominational groups stood a variety of sects and cults described by the census as 'other' Christians. Almost as numerous as the Hindus, they lived in isolated pockets usually associated with low status.

Secondary education remained the hallmark of San Fernando's élite areas (Fig. 4.12). More than one-quarter of the population in southern and central districts achieved this standard, and the proportion exceeded one-third at Spring Vale and St Joseph's Village. At the opposite extreme were Kakatwey and the government housing schemes. The public housing sector was strongly associated with persons who had attained standard 6 or 7 in primary school, but a complete lack of education characterised the town's northern and southern peripheries, and identified Kakatwey as an area of outstanding deprivation, equivalent to Paradise Pasture in 1931.

Linkage analysis

The complex racial characteristics and institutional composition of San Fernando's population may be brought together by statistical analysis of enumeration district data once more using correlation and linkage analysis (Clarke 1973). Eighteen variables have been set out in five groups, which, with the exception of the racial category, are equivalent to the more important institutions specified by M. G. Smith (1965a) in his early formulation of the plural model (Table 4.5). The white population occurred in fewer than three-quarters of the census districts and has been omitted from this part of the study; Catholics and orthodox Protestants have been amalgamated to form

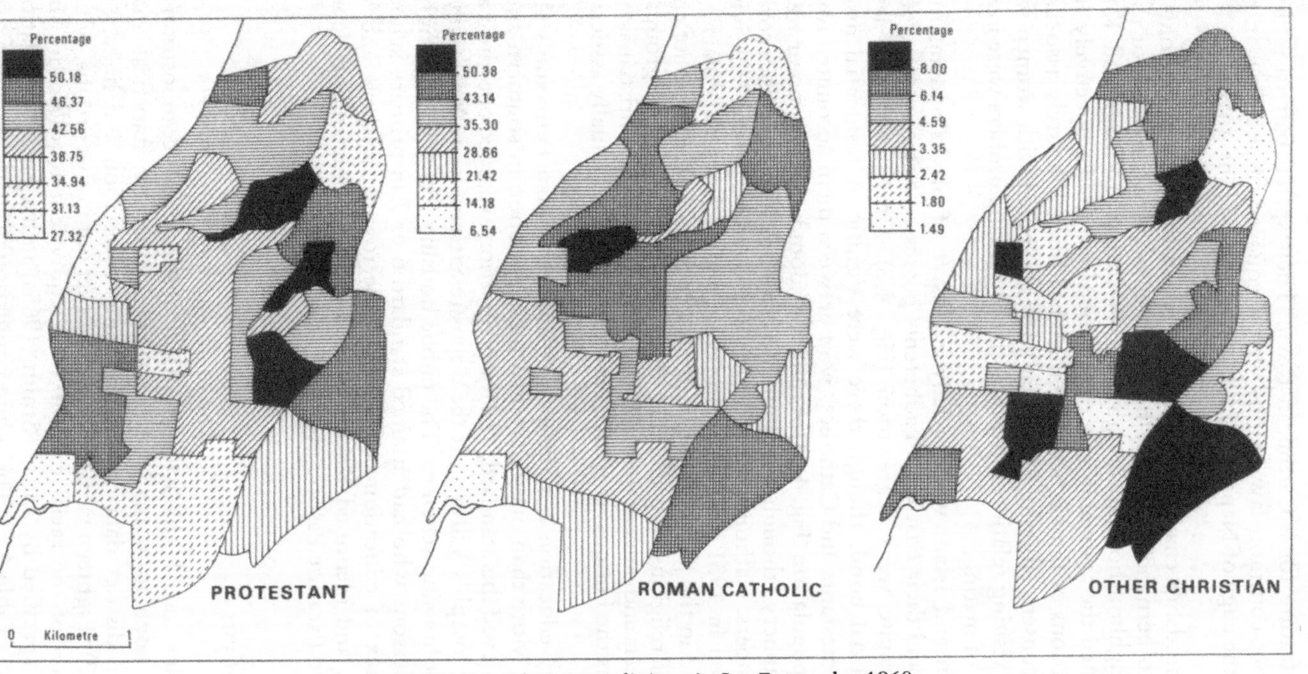

Figure 4.11 Christian religions in San Fernando, 1960.

Table 4.5 Matrix of Spearman rank correlation coefficients for selected variables in San Fernando, 1960 (*source:* Census of Trinidad & Tobago, 1960; —— highest coefficient in each column, ---- second highest coefficient in each column).

Variables	Negro	East Indian	Chinese	Mixed	Denominational Christian	Other Christian	Presbyterian	Hindu	Moslem	Female Head	Female, common law	Female, visiting union	Female, married	Secondary school	Less than standard 6	Standard 6 or 7	No education	Males in non-manual occupations
Negro	1.0	−0.67	−0.41	−0.25	0.62	0.41	−0.47	−0.46	−0.38	0.40	0.28	0.56	−0.34	−0.52	0.30	0.58	−0.35	−0.61
East Indian	−0.67	1.00	−0.25	−0.09	−0.37	−0.16	0.58	0.79	0.50	−0.26	0.05	−0.40	0.20	0.06	0.17	−0.51	0.65	0.16
Chinese	−0.41	−0.25	1.0	0.41	−0.29	−0.19	0.24	−0.02	0.23	0.13	−0.15	−0.25	−0.10	0.43	−0.27	−0.17	−0.26	0.58
mixed	−0.25	−0.09	0.41	1.0	−0.25	−0.15	0.15	−0.35	−0.04	0.26	−0.13	0.11	−0.29	0.42	−0.48	0.23	−0.33	0.53
denominational Christian	0.62	−0.37	−0.20	−0.25	1.0	−0.38	0.05	−0.39	−0.43	0.35	0.19	0.57	−0.17	−0.33	0.27	0.45	−0.12	−0.41
'other' Christian	0.41	−0.16	−0.19	−0.15	−0.38	1.0	−0.06	−0.16	−0.09	0.10	0.32	0.22	−0.19	−0.48	0.25	0.46	−0.06	−0.43
Presbyterian	−0.47	0.58	0.24	0.15	0.05	−0.06	1.0	0.26	0.09	−0.05	−0.13	−0.08	0.31	0.19	−0.06	−0.13	0.31	0.30
Hindu	−0.46	0.79	−0.02	−0.35	−0.39	−0.16	0.26	1.0	0.34	−0.43	−0.01	−0.49	0.29	−0.10	0.15	−0.54	0.67	−0.04
Moslem	−0.38	0.50	0.23	−0.04	−0.43	−0.09	0.09	0.34	1.0	−0.17	0.02	−0.36	0.05	−0.02	0.06	−0.21	0.26	−0.14
Female heads	0.40	−0.26	0.13	0.26	0.35	0.10	−0.05	−0.43	−0.17	1.0	−0.01	0.46	−0.65	−0.02	0.03	0.34	−0.37	−0.09
Female, common law	0.28	0.05	−0.15	−0.13	0.19	0.32	−0.13	−0.01	0.02	−0.01	1.0	0.23	−0.36	−0.44	0.71	0.06	0.37	−0.56
female, visiting union	0.56	−0.40	−0.25	0.11	0.57	0.22	−0.08	−0.49	−0.36	0.46	0.23	1.0	−0.29	−0.32	0.24	0.44	−0.10	−0.40
female, married	−0.34	0.20	−0.10	−0.29	−0.17	−0.19	0.31	0.29	0.05	−0.65	−0.36	−0.29	1.0	0.14	−0.20	−0.12	0.14	0.24
secondary school	−0.52	0.06	0.43	0.42	−0.33	−0.48	0.19	−0.10	−0.02	−0.02	−0.44	−0.32	0.14	1.0	−0.74	−0.39	−0.33	0.88
less than standard 6	0.30	0.17	−0.27	−0.48	0.27	0.25	−0.06	0.15	0.06	0.03	0.71	0.24	−0.20	−0.74	1.0	0.13	−0.49	−0.75
standard 6 or 7	0.58	−0.15	−0.17	0.23	0.45	0.46	−0.13	−0.54	−0.21	0.34	0.06	0.44	−0.12	−0.39	0.13	1.0	−0.40	−0.28
no education	−0.35	0.65	−0.26	−0.33	−0.12	−0.06	0.31	0.67	0.26	−0.37	0.37	−0.10	0.14	−0.33	0.49	−0.40	1.0	−0.29
males in non-manual occupations	−0.61	0.16	0.58	0.53	−0.41	−0.43	0.30	−0.04	−0.14	−0.09	−0.56	−0.40	0.24	0.88	−0.75	−0.28	−0.29	1.0

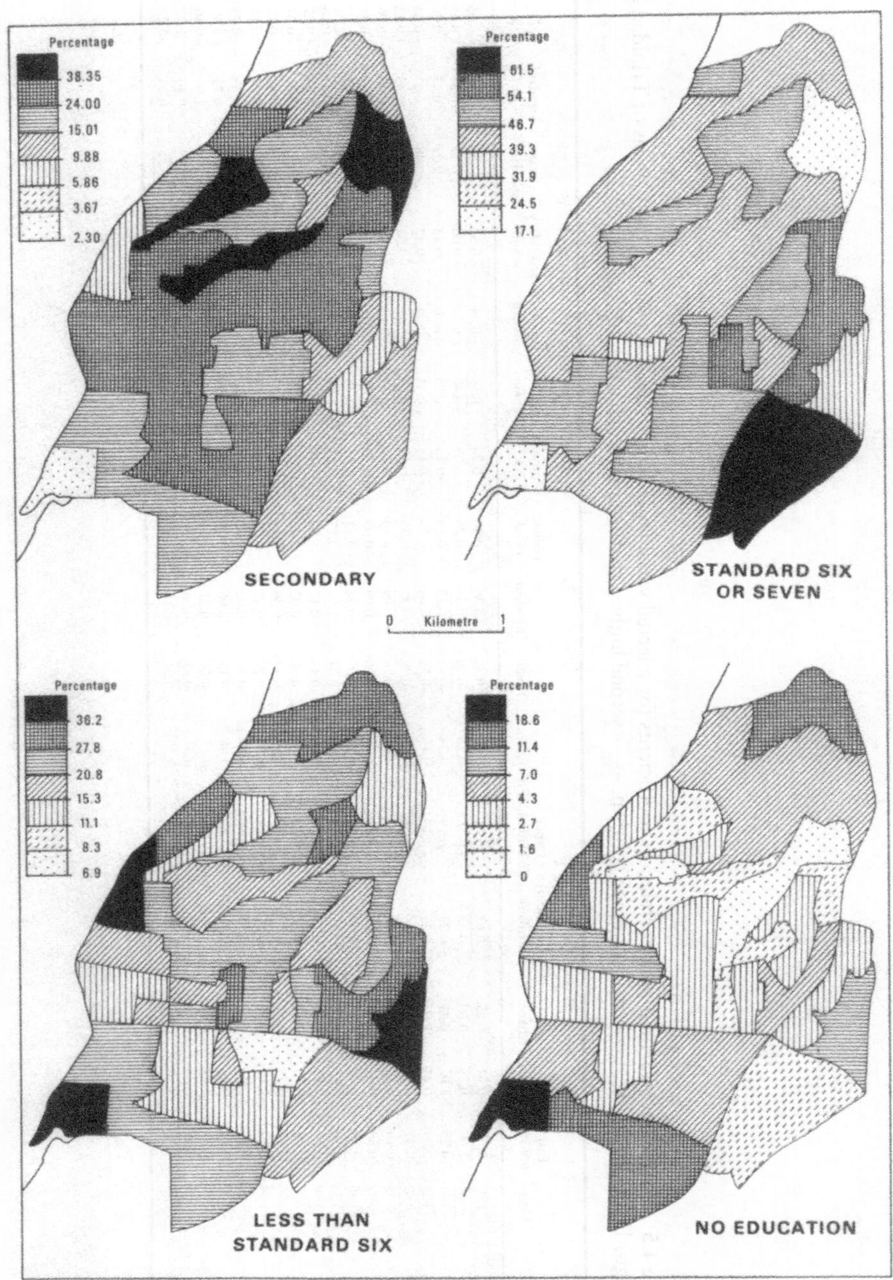

Figure 4.12 Educational standards in San Fernando, 1960.

'denominational Christians'; and occupation has been used as a proxy for Smith's economic institutions. The distinction between manual and non-manual occupations is known to be highly diagnostic of socio-economic variations in the Caribbean.

The linkage diagram (Fig. 4.13; see p. 74) clearly depicts two statistical groups whose composition is reminiscent of the situation in 1931, though change in the census variables for 1931 and 1960 makes detailed comparision impossible. One grouping was based on the strong negative correlation between Negroes and East Indians; the other was structured by the bond between secondary education and males in non-manual occupations, and by the strong negative relationship between these two variables and persons having less than standard 6 education. The first group reflected the most important pluralistic aspects of the urban community; the second expressed the principal features of social stratification. These two groupings were not completely unrelated and were bridged statistically by various aspects of the family.

Between them the two statistical groupings depicted in Figure 4.13 incorporated 17 of the 18 variables used in the analysis. Using the highest and second highest coefficients for each variable, five nodes may be identified. In rank order of the number of correlation bonds associated with them, these nodes are East Indian and non-manual occupations; Negro; less than standard 6 education; and secondary schooling. Examination of these nodes and their satellite variables provides a preliminary survey of San Fernando's social structure in 1960.

Negroes were positively associated with the orthodox Christian churches (0.62), with standard 6 or 7 education (0.58), and with females in visiting unions (0.56); they were negatively correlated with East Indians (−0.67) and Presbyterians (−0.47). As expected, East Indians formed the hub for a series of strong positive bonds with Hindus (0.79), Presbyterians (0.58) and Moslems (0.50). Negroes and East Indians were negatively correlated, as expected, since they were large, mutually exclusive racial categories, and the links between their respective satellite variables were negative too (Fig. 4.13). The variables linked to these two contraposed races included aspects of religion and education, though family differences were apparently of lesser importance. Smith's plural hypothesis is further confirmed by the two top rows of the correlation matrix: 12 of the 16 cases where the Negroes and East Indians were correlated with variables other than one another involved a change of sign when moving from one group to the other. The only four census variables with which both Negroes and East Indians shared the same correlation sign were the Chinese and mixed populations (negative), and females in common-law unions and less than standard six schooling (positive but low).

Males in non-manual occupations were positively correlated with secondary education (0.88) and with the Chinese (0.58) and mixed populations (0.53). They were associated negatively with less than standard 6 schooling (−0.75) and with females in common-law unions (−0.56). Secondary schooling was related positively to non-manual occupations (0.88) and to the Chinese (0.43), and negatively correlated with education below standard 6 (−0.74) and 'other' Christians (−0.48). Finally, education below standard 6 was positively related to females in common-law unions (0.71) and 'other' Christians (0.25), but negatively correlated with males in non-manual occupations (−0.75), with

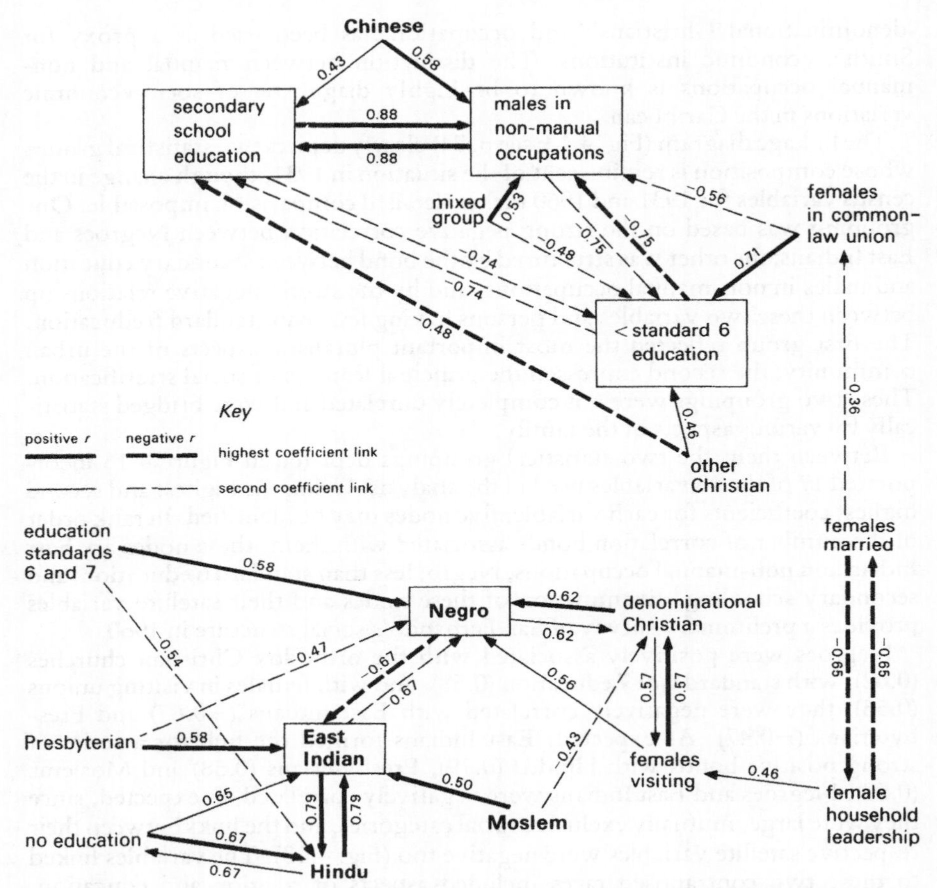

Figure 4.13 Linkage of highest and second highest correlations for each variable in San Fernando, 1960.

secondary schooling (−0.74) and with the mixed population (−0.48). This set of linkages indicates the low status of those persons involved in the Christian cults and sects, or in common-law unions. The mixed population is confirmed as a relatively high status group; so, too, are the Chinese.

Racial groups and their characteristics

Although the linkage analysis is extremely helpful in clarifying the overall pattern of correlation of these variables, it does not, of course, reveal all the most important relationships. Some of these are worthy of further investigation, especially those that converge on race (Table 4.5).

The mixed population recorded positive correlations with males in non-manual occupations (0.53), with secondary schooling (0.42), with the Chinese

(0.41), and, to a lesser extent, with female headship (0.26) and standard 6 and 7 schooling (0.23). Negative correlations for the mixed population were less sharply defined. However, they were even less firmly associated with persons with no education (−0.33), with Hindus (−0.35) and with those having less than standard 6 education (−0.48) than with denominational Christians (−0.25), Negroes (−0.25) and married females (−0.29).

Negroes were characterised as denominational (0.62) or 'other' Christians (0.41); their educational grades were generally standards 6 and 7 (0.58) or less than standard 6 (0.30); and their family structure emphasised the prevalence of visiting unions (0.56), female headship of households (0.40) and the occurrence of common-law unions (0.28). Negroes recorded a negative correlation with non-manual occupations (−0.61) and secondary schooling (−0.52). Moreover, they were negatively related to the East Indians (−0.67), to each East Indian segment (Presbyterians −0.47, Hindus −0.46 and Moslems −0.38) and to the Chinese (−0.41). Comparison of the correlations for the mixed and Negro populations supports the plural hypothesis but shows that the difference between them is far less than between Creoles and East Indians; eight of the 16 cases where the mixed and Negro populations were correlated with other variables involved a change of sign when moving from one racial group to the other. These eight cases of change involved the Chinese, orthodox Christians, 'other' Christians, Presbyterians, females in common-law unions, secondary schooling, education below standard 6 and males in non-manual occupations.

East Indians were associated with Hindus (0.79), Presbyterians (0.58) and Moslems (0.50), and to a lesser extent with the Chinese (0.25) and with married women (0.20). Female headship was an unusual trait (−0.26) and so, too, was membership of the denominational churches (−0.37). Even more atypical of East Indians were visiting relationships (−0.40) and standard 6 or 7 education (−0.51).

The pattern for Hindus was almost identical to that for the entire East Indian group. Moslems, too, recorded similar but lower correlations. Presbyterians deviated little from the East Indian norm, except that they correlated positively with males in non-manual occupations (0.30) and with secondary schooling (0.19). Although Presbyterians represented the most highly acculturated element in the East Indian population, the difference between them and the others was more a question of education and occupation (and religion) than of family structure.

The Chinese were positively linked to variables that expressed status, race and religion. They correlated with males in non-manual occupations (0.58) and secondary schooling (0.43), and showed close relationships to the mixed population (0.41), Presbyterians (0.24) and Moslems (0.23). Negative coefficients are recorded with Negroes (−0.41) and with variables associated with Negroes such as denominational Christians (−0.29), visiting relationships (−0.25), less than standard 6 education (−0.27) and no education (−0.26).

The mixed population has already emerged as a relatively high status group within the Creole segment, and recorded a correlation with non-manual males of 0.53; Negroes formed the base of the Creole stratification (−0.61). East Indians achieved a small positive correlation with non-manual occupations (0.16) and appeared at all levels of employment; Presbyterians (0.30) formed a slightly higher status group, and Hindus (−0.04) and Moslems (−0.14) ranked beneath them. This confirms the approximately parallel ranking of Creoles and East Indians revealed by the indices of dissimilarity. Moreover, the Chinese (0.58) clearly occupied a high status cell within the Creole stratification.

Family structure reflected differences between Negroes and East Indians as well as variations in household composition at different levels in the Creole stratification. The common-law union was associated with low occupational status and poor education, and was more prevalent among Negroes than among other groups. Many lower-class Hindus and Moslems also resorted to consensual cohabitation, but only after their original marriages had foundered. Female headship of households was a Negro phenomenon associated with visiting relationships rather than marital or common-law unions.

Conclusion

Once the 1930s depression was over, the oil industry around San Fernando underpinned the economy of the growing town. Over a 20-year period, spatial growth modified San Fernando's ecology. St Joseph's Village joined Spring Vale as an élite residential area. As the East Indian population rapidly increased, its 1930 core neighbourhood, though still prominent as a Moslem district in 1960, was overtaken by new enclaves around the Hindu temples on the northern and southern peripheries of the town. Coloured Creoles continued to live in close proximity to Negroes, though the residents decanted by slum-clearance schemes from the overcrowded rent-yards and tenements in the town centre gave rise in the 1950s to small black concentrations near the bypass at Navet and Pleasantville.

The Indian-born residents of San Fernando dwindled with the passing of the decades and were not separately enumerated in 1960; nor were *Douglas*. By contrast, the Chinese population increased and stabilised in the commercial centre, where it became prominent in the grocery, restaurant, laundry and photographic trades.

Values placed on topographical features scarcely changed between 1930 and 1960. Elevated ridges and hillsides commanding good views and enjoying sea breezes were favoured by the élite at both dates, whereas the coastal area south of the wharf was the poorest and most rural neighbourhood, with Paradise Pasture being replaced by Kakatwey as the most deprived section of San Fernando.

Even more stable than the spatial pattern was the underlying structure of San Fernando's society. Stability in the indices of dissimilarity from 1931 to 1960 was striking, particularly those for the East Indian religions and those indicating the persistence of white isolation and éliteness. Social mobility among the East Indians enabled them to reduce their social distance from the whites, which had been so marked in 1931, but only marginally affected their relations with coloured Creoles, from whom they were, in fact, not markedly segregated at either date. Likewise, the opening up of élite status to non-whites described in the previous chapter does not seem to have been reflected in the residential mosaic. This may indicate that the residential pattern was slow to adjust to social change: alternatively, the non-white élite areas that were developed made it unnecessary to penetrate white residential space. Indeed, socially mobile East Indians created their own élite neighbourhoods under the influence of racial politics in the late 1950s and early 1960s.

Although the correlation and linkage analysis for 1931 and 1960 cannot be compared in detail, the social structure of the town was similar at both dates and comprised Creole and East Indians plural segments, each internally subdivided. The Creole segment was stratified to form a colour-graded hierarchy of cultures, but among East Indians Presbyterians achieved an occupational status higher than that of Hindus and Moslems, and comparable to that of the Creole mixed group.

This conclusion is consistent with Chapters 2 and 3 – except that the spatial data for San Fernando reveal low segregation between black and brown, Creoles and East Indians, in contrast to the national-level (urban–rural) segregation that existed between them. Whites excepted, low segregation in San Fernando, however, does not imply intermixing or wholesale acculturation, since social isolation of East Indians from Creoles, recently reinforced by political opposition, substantially neutralised the significance of spatial proximity. The principal structures of social differentiation in San Fernando have been touched upon already and have been shown to involve segmentation and stratification, religion and household (Fig 4.13). However, social differentiation can be examined in greater detail only by going beyond the evidence revealed by aggregate census data to consider group and individual behaviour in those very same contexts and with additional reference to intersegmental associations. The following four chapters focus on each of these topics in turn, comparing conditions in San Fernando with those in the village of Débé, located 8 km to the south in the heart of the East Indian rice and sugar district, on the edge of the Oropuche Lagoon.

5 *Segmentation, stratification, race and caste*

The social structure of San Fernando, as revealed by statistical analysis, is both segmented and stratified. Creoles and East Indians constitute co-ordinate segments, each internally stratified by class or occupation, often in conjunction with colour, religion or caste. Race segments East Indians and Creoles; religion reinforces this distinction and further divides East Indians into Hindus, Moslems and Christians; caste is a Hindu phenomenon that has only minor implications for Christian East Indians; shades of colour are important to Creoles but are only of minor significance for East Indians. Among the various aspects of stratification in San Fernando – class, caste and colour – only class is common to all the major racial and religious categories: Creoles, *Douglas*, Christian East Indians, Hindus and Moslems. Yet, as we shall see, social stratification within most categories cannot be reduced to class; nor can segmentation be reduced to race.

Segmentation

East Indians set themselves apart from Creoles by emphasising their racial distinctiveness, their shared Indian past, and their desire to live by the cultural traditions – now greatly modified – which their indentured ancestors brought with them to Trinidad. 'Indianness' is a vital aspect of East Indian cultural identity; in contrast, Creoles are merely 'Western'. Many traits – greetings, clothes, food, artefacts and attitudinal differences – distinguish East Indians from Creoles.

Hindus greet one another with '*Sita Ram*', pressing the palms of their hands together at the same time, as though in prayer. Moslems exchange '*Salaam Alaykum*' and shake hands like Creoles. Traditionally, Hindu men wore a white shirt (*kurta*) and loin cloth (*dhoti*), but these have been given up for Western dress, except on ritual occasions. For many decades women retained the North Indian blouse (*jhula*) and long skirt (*ganghari*), but they have been set aside in favour of dresses with short sleeves, decorated with a long veil or *oronhi*, though even this has been all but abandoned in recent years. Since the mid-1950s, sophisticated East Indian women who think of themselves as 'Indian' have started to wear the *sari*, and more recently the *kurta* and *pyjama*, which are widely used on social and ritual occasions, notably weddings.

East Indians employ a variety of Indian foods and recipes, though many are no longer exclusive to themselves. Rice, *roti*, curry, *dahl*, *channa* and spinach (*bhaji*), all widely prepared staples in rural areas, are a regular part of the East Indian diet in San Fernando. Everyday food is served on enamel or china plates, but Hindu homes usually have brass containers (*lotah* and *tarriah*) for use at religious events (Jha, 1974, 16–17).

Recalling his grandparents' Trinidad village home, Naipaul observed:

> More than in people, India lay about us in things: in a string bed or two, grimy, tattered, no longer serving any function, never repaired because there was no one with this caste skill in Trinidad, yet still permitted to take up room; in plaited straw mats; in innumerable brass vessels; in wooden printing blocks, never used because printed cotton was abundant and cheap and because the secret of the dyes had been forgotten, no dyer being at hand; in books, the sheets large, coarse and brittle, the ink thick and oily; in drums and one ruined harmonium; in brightly coloured pictures of deities on pink lotus or radiant against Himalayan snow; and in all the paraphernalia of the prayer-room: the brass bells and gongs, camphor-burners like Roman lamps, the slender-handled spoon for doling out the consecrated 'nectar' (peasant's nectar: on ordinary days brown sugar and water, with some shreds of the tulsi leaf, sweetened milk on high days), the images, the smooth pebbles, the stick of sandalwood . . . But our community, though seemingly self-contained, was imperfect. Sweepers we had quickly learned to do without. Others supplied the skills of carpenters, masons and cobblers. But we were also without weavers and dyers, workers in brass and makers of string-beds. Many of the things in my grandmother's house were therefore irreplaceable. They were cherished because they came from India, but they continued to be used and no regret attached to their disintegration. It was an Indian attitude, as I was to recognize. Customs are to be maintained because they are felt to be ancient (Naipaul 1964, 31–2).

Faced with the gradual decay of their ancestral culture, East Indians in San Fernando have forged a sense of identity more in opposition to Creole dominance than with reference to their own receding past. In particular they have been concerned less with their own virtues than with their conception of Negro vices – promiscuity, living for the present, fêting. For their part, Creoles have regarded East Indians as financially grasping, litigious, revengeful and deceitful – contriving a surface conformity with what is expected of them while acting in a self-serving manner. As one Christian East Indian noted, 'East Indians are funny: they tell you something, and deep down they mean something else.'

East Indian's attachment to their ancestral homes was much stronger than Creoles'. When asked what areas they would prefer to visit (Appendix B), San Fernando Creoles opted for the USA (48%) followed by Africa (22%). The most popular area among Hindus, Moslems and Débé residents was India (all over 30%), followed in each case by the USA (20%). Only Christian East Indians selected the UK and *Douglas* Canada as their first choice.

India's importance for East Indians was more symbolic than actual. In their sentimental attachment to India, they would abandon their loyalty to West Indian or even to Trinidadian cricket teams when Indian tourists were in the Caribbean. Yet very few East Indians maintained even sporadic contact with kin in India, and most who visited there felt revulsion at the poverty of their ancestral villages.

An important way in which East Indians re-established an awareness of India has been through the cinema. First introduced to Trinidad in 1937, Indian films played a vital rôle in strengthening language skills and a sense of continuity with India at the very period when the last return shipload of ex-indentured Indians set sail for the subcontinent. Indian films, shown in Indian-owned cinemas, projected 'Indianness' rather than sponsoring Hinduism or Islam; they kept alive an image of India's landscapes, and sustained an interest in Indian popular music and mythology, while validating such cultural patterns as arranged marriages. Indian film stars have become heroes and heroines among East Indians, and film music for decades has been an essential accompaniment to East Indian cultural events, particularly weddings.

San Fernando Hindus and East Indians in Débé preferred Indian films to English-language ones: indeed, at Débé, film-going was exclusively associated with the Indian cinema. Creoles and *Douglas*, on the other hand, saw only English-language films; Moslems and Christian East Indians expressed a similar preference. However, almost everyone in San Fernando had at some time seen an English-language film, whereas 65% of Creoles had never seen an Indian film. Film-going markedly polarised urban Creoles and rural East Indians (Hindus, especially), whereas urban East Indians, though maintaining various oriental traits, values and orientations were more attuned to Creole mores.

It would be quite wrong, however, either to think of East Indians as isolated from the main events and pressures of Trinidadian society and economy, or to depict urban East Indians as completely separate from their rural counterparts. An outstanding consequence of the oil boom was that it attracted large numbers of migrants to San Fernando. Over 60% of Creoles, almost 70% of Hindus and more than 50% of Christian East Indians and Moslems were movers. *Douglas* were the least mobile (33%) and most had been literally born and bred in the town, whose racial complexity they reflected. An important aspect of mobility was the high rate recorded at Débé (47.7%). However, this may be the norm for a community where village exogamy (especially among Hindus) prevailed. Indeed, exogamy also helps to account for the very high rate of population movement among Hindus in San Fernando.

More than half the urban Hindu incomers had been born within County Victoria, the head town of which is San Fernando. They were involved in marriages arranged between the children of Hindu townspeople and Hindus in adjacent villages. In this way, conservative rural values and traditions were continually being brought into the town by country-reared girls (the Hindu family is patrilocal) to sustain the community in a culturally complex and creolising environment. Most Christian East Indian, Moslem and *Dougla* migrants, too, had been born in the sugar belt near San Fernando.

Stratification

The Hindu, Moslem and Christian East Indians were racially homogeneous, whereas *Douglas* were of mixed race, and the Creoles and the community at Débé were multiracial. More than 95% at Débé were East Indian; but there were two blacks and three *Douglas* in the sample (see Ch. 1). San Fernando's Creoles were stratified by colour in much the same way as other West Indian populations, with a small, mainly white élite group (2.4%), a median status mixed population (13.8%) and a majority of blacks (82.0%). A small handful of Syrians and Chinese occupied prominent positions based on business activities and the professions in the Creole segment. Local interviewers assessed 9.4% of San Fernando's Creoles as white or fair, 11.4% as light brown, 11.4% as brown and 67.8% as dark brown or black: these data reveal a typical Creole pyramid with a few whites or 'pass as whites', a larger group of brown people and a mass who can be described as black.

More important than colour in San Fernando is occupation, because it differentiates non-Creole as well as Creole. Comparison of the sample populations shows that Creoles and *Douglas* were socio-economically similar (Table 5.1), but Creoles and urban East Indians, although recording a similar range of statuses, were occupationally quite dissimilar (Ch. 4; Harewood 1971). East Indians were prominent in retailing, in road haulage and petrol-station proprietorship, and owned all San Fernando's cinemas. Brown and black Creoles dominated the civil service and the police force. Occupational differences between Hindus and Moslems, and Hindus and Christian East Indians were confirmed as statistically significant by chi-square tests, though Moslems were indistinguishable from Christian East Indians and Creoles. Closer inspection of Table 5.1 shows that more Christian East Indians than other East Indians were white-collar workers; and Hindus were more occupationally polarised than Moslems, with 38% in unskilled or personal service jobs. At Débé, 8% of respondents were agricultural workers, 46% were unskilled or in personal service employment, and a surprisingly high proportion were engaged in skilled

Table 5.1 Relationship between occupational stratification and sample.

Sample	Occupation (no.; as % of population in parentheses)								
	Professional	Manager	White collar	Skilled	Semi-skilled	Unskilled	Farmer	Agricultural worker	Total
Creole	4 (1.9)	22 (10.4)	58 (27.5)	68 (32.2)	27 (12.8)	31 (14.7)	1 (0.5)	0 (0.0)	211 (100.0)
Hindu	5 (3.4)	29 (19.5)	13 (8.7)	22 (14.8)	24 (16.1)	56 (37.6)	0 (0.0)	0 (0.0)	149 (100.0)
Moslem	2 (1.6)	20 (15.8)	35 (27.8)	33 (26.2)	19 (15.1)	16 (12.7)	0 (0.0)	1 (0.8)	126 (100.0)
Christian East Indian	9 (3.5)	36 (14.1)	107 (41.8)	45 (17.6)	32 (12.5)	27 (10.5)	0 (0.0)	0 (0.0)	256 (100.0)
Dougla	1 (2.6)	6 (15.4)	11 (28.2)	9 (23.1)	6 (15.4)	6 (15.4)	0 (0.0)	0 (0.0)	39 (100.0)
Débé	0 (0.0)	5 (4.6)	10 (9.2)	21 (19.3)	14 (12.8)	50 (45.9)	0 (0.0)	9 (8.3)	109 (100.0)

Creole–East Indian: $\chi^2=18.9$, d.f.$=7$, $P<0.01$; Creole–Hindu: $\chi^2=53.6$, d.f.$=6$, $P<0.001$; Creole–Moslem: $\chi^2=5.7$, d.f.$=7$, $P>0.50$; Creole–Christian East Indian: $\chi^2=22.4$, d.f.$=6$, $P<0.01$; Hindu–Moslem: $\chi^2=38.0$, d.f.$=6$, $P<0.001$; Hindu–Christian East Indian: $\chi^2=73.3$, d.f.$=5$, $P<0.001$; Moslem–Christian East Indian: $\chi^2=11.7$, d.f.$=6$, $P>0.05$.

and white-collar work, much of it located outside the village (see also Nevadomsky 1983).

Among Creoles there was still a marked colour–class hierarchy at the beginning of independence. No white was a manual worker and only two out of 14 fair people ranked as low as the skilled-manual grade (Table 5.2). The brown category was split between manual and non-manual occupations, with the light brown majority in non-manual work and the dark brown overwhelmingly in skilled or, to a lesser degree, semi-skilled jobs. However, out of 35 blacks, 33 were manual workers, and browns and light browns dominated the professions (among Creoles). White and fair people concentrated in managerial and other white-collar jobs in the oil industry and commerce.

By contrast, 28 out of 31 persons in personal and unskilled jobs were dark brown or black. Yet the data revealed some signs of upward mobility: although no one darker than brown was in the professions, nine dark brown people were managers, and dark browns formed the backbone of the 'other non-manual' and skilled worker categories (see also Camejo 1971).

San Fernando's colour–class stratification was far from rigid, however, and by 1964 the link between colour and occupation was being further eroded by free secondary education to scholarship winners at 11+, which increased the national availability of free places from 204 in 1955 to 3750 nine years later (Williams 1969, 318). Moreover, hostility among people of different shades of colour, so common in Caribbean Creole societies, had for long been muted in Trinidad. Despite the long history of white éliteness in San Fernando, anti-white feeling was absent at independence. This was partly because Creoles realised they must sink their differences to overcome East Indian opposition, but also because throughout the decades of Indian immigration the black, brown and white Creole populations had developed a sense of community in opposition to the East Indians.

Educational patterns inherited from the past, when scholarships were few and secondary education had to be paid for, paralleled differences in employment (Cross & Schwartzbaum 1969). Creoles and East Indians were significantly different from one another, in terms of education, yet variations within the East

Table 5.2 Relationship between colour and occupation for Creoles (sample).

		Occupation							
Colour	Total	Profes-sional	Manager	Other non-manual	Skilled worker	Semi-skilled	Personal services unskilled	Farmer	Agricultural worker
white	6	0	4	1	0	0	0	1	0
fair	14	0	5	7	2	0	0	0	0
light brown	24	2	2	11	4	2	3	0	0
brown	24	2	1	9	10	2	0	0	0
dark brown	108	0	9	29	41	14	15	0	0
black	35	0	1	1	11	9	13	0	0
Total	211	4	22	58	68	27	31	1	0

Plate 1 East Indian retailers, High Street, San Fernando.

Plate 2 Hindu cremation, the Creek, San Fernando. The Gulf of Paria is in the background.

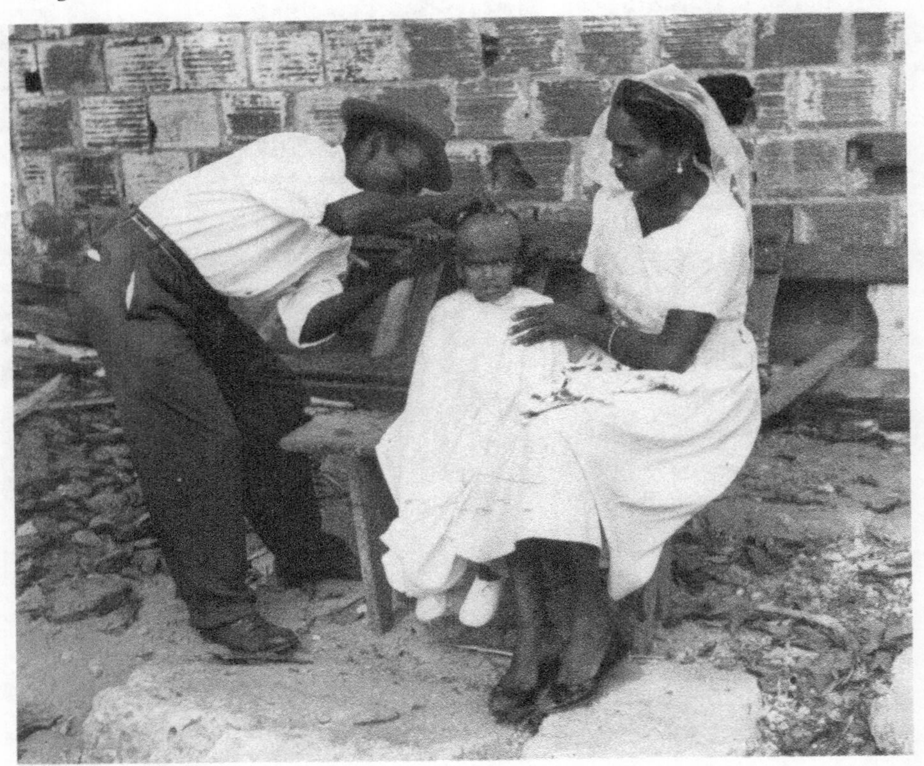

Plate 3 First cutting of a boy's hair, *Siparu Mai* celebration, Siparia. The woman is wearing an *oronhi* (veil).

Indian segments were almost as pronounced (Table 5.3). Christians were the best educated East Indians, Hindus the worst: more than 16% of Hindus in San Fernando had had no formal education whatsoever, a proportion that was even higher than for Débé, where only a small minority (6.5%) had been to secondary school.

Except among Hindus, of whom 20% were illiterate, virtually everyone in San Fernando could read and write English. Facility in Hindi or Urdu was almost entirely restricted to East Indians: the majority of East Indians – even Christians – claimed to comprehend spoken Hindi. Over 20% of Hindus in San Fernando and Débé, and even a handful of Christian and Moslem East Indians, could read and write Hindi or Urdu.

Despite marked differences in occupation, educational achievement and language facility between Creoles and East Indians, respondents from both segments had highly similar aspirations for their sons and daughters. All were extremely ambitious (see also Rubin 1959, Green 1965, Rubin & Zavalloni 1969, Baksh 1979). Almost everyone wanted their offspring to go to university, though urban Hindus had their sights set slightly lower.

More than two-thirds of all those sampled wanted their sons to enter professional employment; this included respondents at Débé, though Creole aspirations were a little lower than those of East Indians. No one mentioned manual tasks as desirable future occupations. Parents were also very ambitious for their daughters, with the professions topping a list that included teaching and nursing. East Indians (especially Hindus) expressed a stronger preference for professional work than Creoles, who also stressed nursing as did people at Débé. However, the homogeneity of respondents' ambitions masked considerable discrepancy between aspirations and likely achievements in the short run.

Table 5.3 Relationship between educational attainment and sample (no.; as % of population in parentheses).

Sample	No education	Standard not given	Kindergarten standard 3	Standard 4 or 5	Standard 6 or 7	No school cert.	School cert.	With degree	Without degree	Total
Creole	1 (0.5)	1 (0.5)	8 (3.8)	24 (11.4)	110 (52.1)	37 (17.5)	28 (13.3)	2 (0.9)	0 (0.0)	211 (100.0)
Hindu	25 (16.8)	3 (2.0)	29 (19.5)	27 (18.1)	33 (22.1)	17 (11.4)	13 (8.7)	2 (1.3)	0 (0.0)	149 (100.0)
Moslem	6 (4.8)	0 (0.0)	12 (9.5)	21 (16.7)	49 (38.9)	23 (18.3)	13 (10.3)	2 (1.6)	0 (0.0)	126 (100.0)
Christian East Indian	6 (2.3)	0 (0.0)	15 (5.9)	23 (9.0)	113 (44.1)	49 (19.1)	42 (16.4)	6 (2.3)	2 (0.8)	256 (100.0)
Dougla	1 (2.6)	0 (0.0)	0 (0.0)	0 (0.0)	22 (56.4)	5 (12.8)	9 (23.1)	2 (5.1)	0 (0.0)	39 (100.0)
Débé	16 (14.7)	3 (2.8)	21 (19.3)	21 (19.3)	41 (37.6)	4 (3.7)	3 (2.8)	0 (0.0)	0 (0.0)	109 (100.0)

Creole–East Indian: $\chi^2=30.7$, d.f.$=7$, $P<0.001$; Creole–Hindu: $\chi^2=81.3$, d.f.$=7$, $P<0.001$; Creole–Moslem: $\chi^2=17.4$, d.f.$=7$, $P<0.02$; Creole–Christian East Indian: $\chi^2=9.5$, d.f.$=7$, $P>0.20$; Hindu–Moslem: $\chi^2=24.7$, d.f.$=7$, $P<0.001$; Hindu–Christian East Indian: $\chi^2=74.1$, d.f.$=7$, $P<0.001$; Moslem–Christian East Indian: $\chi^2=10.7$, d.f.$=6$, $P>0.05$.

Aspirations of all groups soared way above the capacity of the local economy and educational system to supply what they desired: even after the PNM's reforms, barely 20% of the relevant age group attended state secondary schools at 11+, so access to high educational and occupational status during the next two decades seemed likely to remain severely restricted.

In addition to other information collected about all persons in each sample, questions were asked about the household in which the respondent resided. These questions referred to the economic characteristics of the household head – focusing on employment and social mobility – and the fabric and quality of the house and its key contents.

Petroleum provided most of San Fernando's industrial employment. The industry directly employed between 10 and 18% of adults in each urban sample, the highest proportions recorded being for Creoles, Moslems and Christian East Indians. Although there was a small industrial estate on the northern outskirts, manufacturing was of minor importance as an employer of labour; among East Indians it was completely outstripped by transport, storage and communications, and in all samples it was overshadowed by commerce and construction. The dearth of manufacturing and inflation of the 'other services' sector (mostly personal and domestic service), which accounted for the employment of at least a quarter of the adults (and a higher proportion of women) in each urban segment, indicated how far the health of San Fernando's economy depended on the oil industry. Of course, the division of the data into discrete categories misrepresents the complexity of the job situation, especially among the lower classes, where occupational multiplicity was very common. One East Indian encountered during the research claimed to be employed in six different households; and many Creole men were known to have two or three jobs, a common supplementary activity being taxi driving (Thomas 1972).

Movement into the élite was examined in Chapter 3; but did economic change in San Fernando create avenues for more widespread social rising? This can be answered by comparing the proportions of male household heads in white-collar work with their fathers. Only one group displayed retrogression: 46.4% of *Dougla* fathers had been engaged in white-collar work, but only 35.8% of their children. Many of the *Douglas* in the sample were the offspring of sophisticated East Indian fathers who moved in racially mixed social circles. All other samples moved occupationally upward, notably the East Indians, including the Débé sample; Christian East Indian heads recorded the largest white-collar proportion (49.5%), but intergenerational mobility was even greater among urban Hindus. The survey confirms the analysis of change among the élite described in Chapter 3 and shows it to have applied as a more general social principle.

The picture of San Fernando which emerges is of a prosperous oil-based economy diversified by commerce, transport, services and construction, providing ample scope for social mobility. In 1946, fewer than 3% of the local labour force had been unemployed compared to 6% for Trinidad as a whole (Farrell 1978), and the 1964 survey revealed still fewer unemployed: those wanting work but getting none ranged from 0.9% of Creole household heads to 4.5% of Hindus – in Débé the rate was zero. Admittedly, underemployment was more common than outright unemployment, with 31.8% of Creole heads working for four or fewer days in the week prior to the survey, compared to

24.5% for Hindus and Christian East Indians, 21.3% for Moslems and 17.9% for *Douglas*. Where unemployment was lowest, underemployment was highest, and this was well exemplified at Débé where 31.8% of household heads worked for four or fewer days per week; and it was presumably in the context of underemployment that one Creole respondent remarked, 'There are more unemployed Negroes than East Indians, as the Negroes do not do any jobs they consider degrading, such as scavenging and labouring on the roads.'

Although the data on occupational differentiation suggest that San Fernando's major segments were internally class-stratified, the information on income confirms that each segment enjoyed similar levels of material wellbeing. The best-off household heads, earning an annual income of more than $5000 (Trinidad & Tobago; £1000), included 15% of the Creoles, 15% of the Hindus, 13% of the Moslems, 20% of the Christian East Indians and 20% of the *Douglas*. This illustrates East Indian socio-economic mobility in the course of the 20th century, and confirms the progress of non-Christian East Indians via trade in the 1940s and 1950s. Virtually no household head had inherited wealth, but profits and fees were notably Hindu and Moslem features, whereas salaries and wages typified Creoles, Christian East Indians and *Douglas*. Marginality among the Creole poor was revealed by their reliance on private and public relief. Rural East Indians at Débé were characterised by two major sources of income – profits and fees (29.4% of household heads) and wages (54.1%); but not a single household head in Débé received an income in excess of $5000 (Trinidad & Tobago), and rural East Indians were clearly much worse off than the urban segments (Dookeran 1974).

The survey of heads of households and respondents confirms the aggregate 1960 census data on occupations: differences in employment structure between Creoles and East Indians in San Fernando were comparatively slight; all the major samples were internally class-stratified and contained a similar range of occupational categories. However, the samples differed quite markedly in their priorities for cash expenditure, and this is exemplified by considering the most costly outlay facing households – acquisition of a home. The sociopsychological literature on Trinidad suggests that Creoles are orientated towards 'living for the present', whereas East Indians opt for 'delayed gratification' and make long term financial and educational plans (Mischel 1961; Rubin & Zavalloni 1969). House tenure in San Fernando was compatible with this distinction: Hindus, Moslems and to a lesser extent Christian East Indians generally owned their homes; Creoles and *Douglas* rented. East Indian preoccupation with property and house ownership, so beautifully depicted by V. S. Naipaul (1961) in *A house for Mr Biswas*, was at work even in Débé, where no one rented. Only under conditions of extreme urban poverty did East Indians slip into squatting, as at Kakatwey.

Ownership of reinforced concrete homes was common among all East Indians; Creole homes, like the rustic structures at Débé, tended to have wooden walls. However, otherwise Creoles and East Indians were differentiated by housing materials no more than by occupation: every urban sample recorded houses of all grades, though urban Creoles and Hindus were more strongly associated with poor and very poor dwellings than the others. Property at Débé clustered in the categories average and fair.

House tenure distinguished between East Indians and Creoles, but ownership of consumer durables was common only among Moslems and Christian East Indians. Cheap items such as radios and rediffusion sets were almost ubiquitous; only at Débé did fewer than 29% of households have them. Conversely, expensive luxuries, such as televisions were only just being acquired in San Fernando in 1964 – especially by Christian East Indians and Moslems – and had barely reached rural areas such as Débé. Ownership of cars, refrigerators and houses with piped water and water closets, were discriminant indicators, however. Moslems led in the possession of all these prestigious items, and urban Hindus and East Indians in Débé had fewest, with Creoles and *Douglas* in intermediate positions. Hindus conformed to the 'Indian' cultural trait of home ownership but could not, in addition, afford sophisticated household effects enjoyed by Christian East Indians and Moslems, especially the 37.8% of Hindu household heads who were unskilled or personal service workers.

These household data generally confirm the ecological data set out in Chapter 4. Residential segregation in San Fernando was low because the housing market was open and each segment was internally differentiated by class. Each sample contained a range of property of different quality, but East Indian social mobility and preoccupation with house ownership led East Indian realtors to develop new neighbourhoods. Conversely, black concentrations remained in the public rental sector.

Race

Although Creoles and East Indians shared many goals including similar aspirations for their children, they were keenly aware that racial ascription often constrained availability of job and other opportunities. Indeed, this may help to account for their overwhelming preference for professional careers for their children, which conferred prestige and high income without requiring public and private employment.

All East Indians agreed that it was difficult for Negroes to get jobs in banks, and one young Creole observed, 'Negroes are not loved by the whites; they prefer the East Indians. It is a waste of time for Negroes to apply for jobs in the banks.' All samples perceived bank employment as being more accessible to East Indians, except the inhabitants of Débé who envisaged themselves as barred (perhaps on educational grounds). In reality, only phenotypically fair people worked in San Fernando's banks in 1964; some Chinese Creoles and light-skinned East Indians were tellers.

Most Creoles agreed it was difficult for Negroes to get employment with East Indians (as did the Débé sample), but all urban East Indians denied it, and so, too, did the *Douglas*. Conversely, many East Indians found it was difficult for them to get jobs with Negro employers, this in turn being denied by Creoles and *Douglas*. Respondents in each segment agreed it was more difficult for Negroes than East Indians to get employment in Chinese enterprises, but in each case, except Débé, the percentages were small. Urban and rural East Indians alike believed they had easier access than blacks to jobs in white-owned businesses, but Creoles and *Douglas* saw no racial difference in job chances.

Several caveats are essential at this point: East Indians employed family labour, as did the Chinese; white business houses tended to be multiracial but organised hierarchically, as exemplified in Mittelholzer's (1964) *A morning at the office*; and one respondent commented 'While the Chinese will employ Negroes in their businesses, it is only the labouring jobs they will be given to do.'

Creole–East Indian attitudes toward public-sector employment were even more polarised, for here racial enclaves were evident. The police force in San Fernando in 1964 was substantially Creole, as too was borough government: only 11% of Trinidad's civil servants and 2.5% of its police were East Indian (Lowenthal 1972, 167–8). However, in the public health office and the probationary service for County Victoria, where department heads were East Indian, the staff, too, were almost entirely of the same race.

Virtually no respondent in the survey thought the police force fairly represented the major racial categories, though a few in all samples ascribed East Indian under-representation to failure of the entrance tests, with Creoles, *Douglas* and the Débé community underlining this point (Table 5.4). However, the most popular explanations for the dearth of East Indians in the police force were quite different. East Indians attributed their absence to racial discrimination or to their own dislike of the work; Creoles and *Douglas* discounted racial discrimination but emphasised that police work was distasteful to East Indians (Table 5.4).

In certain respects police work was indeed antipathetic to many East Indians – especially to Hindus and Moslems – because of their dietary taboos and reluctance to live in barracks in close proximity to Creoles. However, no such reservations attached to the civil service, which was well paid, secure and respectable. Almost one-third of Creoles and *Douglas* denied there was racial inequality in the civil service; even East Indian respondents were less adamant

Table 5.4 Why are there proportionately fewer East Indians than blacks in the police? (No.; as % of sample in parentheses.)

Sample	Race pressure	Fail entrance test	Don't want job	Food problems	Don't know	Fail physical	No inequality	Total
Creole	6 (2.8)	53 (25.1)	107 (50.7)	1 (0.5)	11 (5.2)	33 (15.6)	0 (0.0)	211 (100.0)
Hindu	56 (37.6)	23 (15.4)	49 (32.9)	1 (0.7)	10 (6.7)	9 (6.0)	1 (0.7)	149 (100.0)
Moslem	33 (26.2)	25 (19.8)	48 (38.1)	1 (0.8)	4 (3.2)	15 (11.9)	0 (0.0)	126 (100.0)
Christian East Indian	79 (30.9)	50 (19.5)	97 (37.9)	0 (0.0)	7 (2.7)	21 (8.2)	2 (0.8)	256 (100.0)
Dougla	3 (7.7)	10 (25.6)	20 (51.3)	0 (0.0)	0 (0.0)	6 (15.4)	0 (0.0)	39 (100.0)
Débé	45 (41.5)	26 (23.9)	25 (22.9)	3 (2.8)	6 (5.5)	4 (3.7)	0 (0.0)	109 (100.0)

Table 5.5 Why are there proportionately fewer East Indians than blacks in the civil service? (No.; as % of sample in parentheses.)

Sample	Racial pressure	Fail entrance	Don't want job	Don't know	No inequality	No reply	Total
Creole	8 (3.8)	24 (11.4)	45 (21.3)	65 (30.8)	69 (32.7)	0 (0.0)	211 (100.0)
Hindu	94 (63.1)	15 (10.1)	11 (7.4)	21 (14.1)	6 (4.0)	2 (1.3)	149 (100.0)
Moslem	75 (59.5)	25 (19.8)	9 (7.1)	10 (7.9)	7 (5.6)	0 (0.0)	126 (100.0)
Christian East Indian	166 (64.8)	31 (12.1)	22 (8.6)	19 (7.4)	18 (7.0)	0 (0.0)	256 (100.0)
Dougla	10 (25.6)	6 (15.4)	8 (20.5)	3 (7.7)	12 (30.8)	0 (0.0)	39 (100.0)
Débé	72 (66.1)	16 (14.7)	8 (7.3)	12 (11.0)	1 (0.9)	0 (0.0)	109 (100.0)

Creole–East Indian: $\chi^2=261.9$, d.f.=4, $P<0{\cdot}001$; Creole–Hindu: $\chi^2=163.4$, d.f.=4, $P<0.001$; Creole–Moslem: $\chi^2=157.6$, d.f.=4, $P<0.001$; Creole–Christian East Indian: $\chi^2=204.0$, d.f.=4, $P<0.001$; Hindu–Moslem: $\chi^2=7.7$, d.f.=4, $P>0.10$; Hindu–Christian East Indian: $\chi^2=7.0$, d.f.=4, $P>0.10$; Moslem–Christian East Indian: $\chi^2=11.1$, d.f.=4, $P<0.05$.

about their under-representation than in the case of the police (Table 5.5). Most East Indians, irrespective of segment attributed their under-representation to racial discrimination, whereas those Creoles who sought to account for racial inequality said the East Indians didn't want civil-service work.

Differences between Creoles and East Indians over perceived access to employment were not based upon theoretical assumptions about innate racial superiority or inferiority. Only about 10% of the East Indians affirmed the racial superiority of any group, the percentage dropping to 7.7 for *Douglas* and 2.4 for Creoles. Those who did inevitably picked out whites as the master race and denigrated blacks. A Moslem respondent typified several others in observing, 'White people and Indians should rule, and Negroes should be servants, for according to the ancients the latter have received a curse from above. Negroes are the lowest of races.'

Most Creole and East Indian respondents believed that they got on 'fairly well', *Douglas* and Débé residents expressing the greatest reservation. All East Indians living in San Fernando were apt, like a Moslem respondent, to attribute bad relations to 'a racial feeling from the government'. Only Hindus showed any inclination for segregation (45.9% of the Débé sample being favourably disposed – perhaps because they had already achieved it); Creoles, *Douglas* and Christian East Indians were opposed to it. By contrast, Creoles and *Douglas* were strongly approving of miscegenation (87%), whereas East Indians, especially Hindus, were overwhelmingly opposed to it.

What was the experience of *Douglas* in the polarised social circumstances of San Fernando? Did they provide a bridging element; or did they gravitate towards the Creoles, often entering the coloured middle class as the sample respondents suggest? The life history of one *Dougla* informant conforms to the latter pattern. Fathered by a member of a prominent East Indian family in San Fernando, he was raised in another town by his Spanish-Creole mother. When he left Catholic secondary school, he worked for a while on his father's small coconut plantation before doing odd jobs and going to sea. Later he got married, settled in San Fernando, and had three children, whom, at the time of the interview, he was supporting by van driving and part-time taxi work using his own car. Claiming that his East Indian forbears had rejected him because he was illegitimate and half Negro, he was orientated to the Creole segment. His wife was black, and his casual affairs invariably involved black girls.

Caste

Evidence concerning the experience of the *Douglas* confirms that East Indians expel their coloured offsping into the Creole population, where racial mixing has been a continued process; the East Indians thereby ensure racial exclusiveness. Yet caste is one of the few basic elements of Indian culture that Hindus say they have not preserved. Nevertheless, most urban as well as rural Hindus know their caste affiliation; caste labels persist among many Christian East Indians; and caste so intersects with class among urban Hindus that East Indian stratification cannot be understood without attention to it.

East Indians of Hindu origin in Trinidad think in terms of high, medium and low castes, which correspond to the varna system of India. Varnas are ranked hierarchically: the highest is the Brahmin, followed by the Kshatriya, Vaishya and Sudra. No outcastes are recognised by Hindus in Trinidad, for everyone was polluted and outcaste by crossing the sea or *kala pani* (black water), and in Trinidad traditional Indian outcastes have been incorporated into the Sudra varna. However, folk ideas about varna have been kept alive by oral tradition, reinforced by Hindu texts, such as the *Bhagavad-gita*.

Caste labelling has been retained at Débé (Niehoff 1967, 153) and to a lesser extent in San Fernando; 20 Hindus (13% of the San Fernando sample) and 144 Christian East Indians (60%, omitting those of non-Hindu origin) did not know their caste (*jat*), compared with only two at Débé (2%, when non-Hindus are set aside). However, conversion to Christianity has led to a loss of caste knowledge – perhaps deliberately among members of low castes. Thus only 3% of Christian East Indians confessed to being Sudra, compared with 7% among San Fernando Hindus and 31% among Hindus at Débé. A Presbyterian shopkeeper expressed the common Christian opinion when he claimed not to know either his own or his wife's caste. He added that he was fed up with East Indians in Trinidad still thinking about caste: East Indians should all become Christians 'and so change their outlook on life'.

The majority of Trinidadian rural Hindus were Vaishya or Sudra, with fewer in the high Kshatriya and fewer still in the Brahmin varnas (Table 5.6, Klass 1961, 61, Schwartz 1967a). Among Hindus in San Fernando, however, the

Table 5.6 The caste of samples of Hindus and Christian East Indians in San Fernando and of Hindus in Débé (respondent and spouse) (*source:* Amity data from Klass (1961, 61)).

Caste and varna	Traditional occupation of castes	San Fernando Hindus Male	San Fernando Hindus Female	San Fernando Hindus Total	San Fernando Christian Total	Débé Hindu total	Amity (total population; major castes) total
Brahmin							
Gosain		3	1	4	1	3	
Maharaj	priests	32	22	54	29	12	23
subtotal		35	23	58	30	15	23
Kshatriya							
Chattri	warriors and rulers	22	22	44	45	33	54
subtotal		22	22	44	45	33	54
Vaishya							
Ahir	graziers	20	13	33	18	22	68
Baniya	traders	3	3	6	9	5	18
Barhai	carpenters	0	2	2		1	
Gadariya	shepherds, goatherds and blanket weavers	1	1	2			
Jaiswal	traders	1	0	1			
Kahar	cultivators, fishermen and carriers of palanquins	3	1	4	3	2	25
Kewat	fishermen, boatmen and cultivators	1	0	1	2	2	
Koiri	cultivators	4	3	7	3	4	36
Kori	weavers	0	1	1			
Kurmi	market gardeners	10	8	18	8	13	16
Lunia	cultivators and makers of earthworks	3	1	4	4	1	
Madrassi	name given to people from South India (see text)	2	0	2	5	2	11
Mali	gardeners	0	1	1			
Mallah	fishing and boating	1	0	1	1		14
Nau	barbers	1	2	3	1	3	9
Patitar		1	0	1			
Sonar	goldsmiths	0	1	1	2		
Bhujawa							10
Banya							31
Teli	oil pressers, traders and cultivators	2	0	2			
Vesh		1	1	2			
Bori							30
Murao	cultivators				2		
Lohar	blacksmiths				1		
Kayat	clerks				1		
Lalla	unknown				1		
Dhobi	washermen				2		
Bind	labourers and ploughmen				1		
subtotal		54	38	92	64	55	268
Sudra							
Dusad	ploughmen and village menials	6	5	11	2	7	21
Seunerine	(see text)	1	0	1			
Chamar	tanners and leather workers	5	11	16	3	40	282
Bhar	ploughmen and day-labourers				1	1	68
Parsi	collectors of palm oil and day-labourers					1	12
Dom							28
subtotal		12	16	28	6	49	411
Dass		0	1	1	0	0	0
Total		123	100	223	145	152	756

Sudra varna was truncated, the proportions of Kshatriyas and Vaishyas were as expected, and Brahmins greatly over-represented (Table 5.6). Among Christian East Indians, Brahmins slightly exceeded the Débé percentage, but all other varnas were under-represented because of loss of information or deliberate amnesia.

Caste among Hindus in San Fernando Each Hindu respondent in San Fernando was asked to state his or her caste and the caste of his wife or her husband. The caste of 223 adults – 123 men and 100 women – was recorded and among them 26 castes were listed (Table 5.6): only three respondents mentioned their subcaste.

The Brahmin or priestly varna in San Fernando comprised two castes, Gosain and Maharaj. The Gosain caste was very small, but it was agreed among the Brahmins that it was the highest *jat* or 'nation'. Many Brahmins have adopted as a surname the name Maharaj or 'great ruler': they claimed membership of the Maharaj caste but most did not know their subcaste. However, the names of some Brahmin castes have been retained as family names – notably the Misir, and the Dube.

The only caste in the Kshatriya varna, the varna of rulers and warriors, was called Chattri; it contained no subcastes.

The third, or Vaishya, varna comprised 19 castes. In India members of most of these castes had been involved in trading, agriculture, fishing and rural crafts. In the source areas in North India from which indentured Indians came the Kewat caste was sometimes listed as a subcaste of the Mallah; the Jaiswal was associated with the Baniya; and the Mali, Koiri and Kurmi were closely aligned (Crooke 1896). Castes were ranked within the Vaishya varna in at least one rural area of Trinidad (Klass 1961), but the breakdown of caste feeling within the varna in San Fernando made the hierarchy inconsequential there and the Vaishya castes have therefore been listed alphabetically.

Three castes – Dusad, Seunerine and Chamar – comprised San Fernando's Sudra varna. In India the Dusad were traditionally ploughmen, watchmen and village menials; the Chamar were tanners, leather workers and day-labourers; and the Chamarin (female members of the Chamar caste) were midwives – a polluting task. The Chamar were despised and ostracised throughout North India because they worked with the skin of dead animals. There is no Indian caste called Seunerine, though there is a religious sect (Siva Narayani) of that name. The sect drew the greater part of its support from the Sudras, and I have placed the Seunerine in the Sudra varna.

In India varna is associated with race or skin colour: the higher castes are light, the lower castes dark. This link is still expected in San Fernando; an oft-repeated saying warns against the treachery of people whose colour and caste do not coincide in this way: 'Beware of the black Brahmin and the fair Chamar.' In San Fernando, other Hindus accepted the pre-eminence of the Brahmins because of their traditional ritual status, and measured the rank of other castes by their proximity to Brahmins. However, social rising and demotion have also been experienced by individuals. In San Fernando, a Maharaj who ate pork was a 'pork Maharaj', and because of his pollution was considered lower even than a Chamar. However, as class is a more important common denominator than caste in San Fernando, members of the low castes have more frequently sought to upgrade themselves by acquiring educational credentials and white-collar work than by adopting orthodox Hindu behaviour.

A good example of local change in caste position is provided by the Madrassis. As discussed in Chapter 2, Madrassi is an umbrella term applied to the descendants of indentured labourers shipped from South India through the port of Madras. Their customs and rituals differed from those of the majority of Kalkatiya Indians, who considered all Madrassis as lower than Chamars. However, some Madrassis in San Fernando countered that they comprised two strata: the Moon-Sammies, who were Brahmins, and the Mootoos, who were Sudra. Neither of the two Madrassis interviewed in the survey mentioned this distinction, however, and the Madrassi caste in San Fernando ranks as Vaishya.

The striking features of San Fernando caste make-up were the numerical preponderance of the three twice-born varnas over the Sudras; the fact that Brahmins and Kshatriyas together outnumbered the Vaishyas; and that Brahmins alone also outnumbered the Kshatriyas (Table 5.6). Varna size in San Fernando reversed the varna proportions of the original immigrants, and also the proportions recently recorded for the Trinidad village of Amity (Klass 1961, 61), where Klass found over half the males to be Sudra, and well over half the Sudra, Chamar. In San Fernando, also, over half the Sudra were Chamar, but only 12.5% of Hindus were Sudra.

Over-representation of the higher castes in San Fernando is explicable by differential social, economic and geographical mobility. During the early decades of the 20th century, Brahmin priests were among the wealthiest and most influential members of the Hindu community; Brahmins were self-confident and ambitious and more likely than the Sudra to seek opportunities for material advancement in urban schools and jobs. Many low-caste members of the Hindu community who were born or came to live in San Fernando probably converted to Christianity: the Presbyterian Church provided both a way out of the canefields – and out of caste.

In addition to urbanisation of the higher castes and conversion of the lower ones, falsification of caste may also have occurred. It is suspected that some lower caste Hindus claimed Brahmin status when they came to Trinidad. Ridicule and beatings were the lot of those who were found out, and this must have been a deterrent to the majority. Nevertheless, two non-Brahmin men were known to have claimed and been granted Brahmin status in San Fernando: one described as a 'half-way' Moslem, the other the son of a Kshatriya. The latter was a politically important figure; his own wealth and his father's high caste made him acceptable to all Brahmins. In short, indentured immigration followed by cityward migration enabled some Hindus to 'adjust' their caste affiliation. However, the long inventory of urban Vaishyas and of Sudras shows this did not happen on a large scale. Certainly the caste of most Brahmins was undisputed; and several had pedigrees based on ancestors who had been well-known Hindu priests.

Caste and class

Did even those San Fernando Hindus who knew their caste affiliations practise caste as a system? If we accept Hutton's definition of a caste system as 'one whereby a society is divided up into a number of self-contained and completely segregated units (castes), the mutual relations between which are ritually determined in a graded scale' (Hutton 1961, 50), it is clear that they did not. A sketch of 19th century caste abandonment has already been given (Ch. 2). By 1964, caste was an issue for the individual or family, but not for any larger organised group. With the exception of the orthodox priesthood, still confined to Brahmins, Hindu occupations no longer bore any of the caste linkages traditional in 19th-century India. The dissolution of caste in its Indian sense is beautifully summarised by V. S. Naipaul:

> In Trinidad caste had no meaning in our day-to-day life; the caste we occasionally played at was no more than an acknowledgement of latent qualities; the assurance it offered was such as might have been offered by a palmist or a reader of handwriting. (Naipaul 1964, 36).

To most San Fernando Hindus, therefore, caste presented no great problem. In their working lives race, religion and class were infinitely more important than caste. Many Hindus claimed that caste did not matter; some denied its existence. However, caste did matter at the level of stereotype and rhetoric, as the following quotations from respondents and informants show:

> *Ahir, aged 28.* The caste system is not the best thing, but I'm in it and what can I do? Quicker trust a Maharaj than a Chamar. The Chamar act and think low.
>
> *Chamar, aged 48.* The only hope for Hindus who are not Maharaj is to become Christians, when their status in life will become changed in the eyes of these Brahmins.
>
> *An Indian-born Brahmin school teacher.* Caste is silently felt, especially in the upper bracket . . . No one is diffident about caste in Trinidad, but many are arrogant . . . The high castes are arrogant.

Brahmins and Kshatriyas were proud of their caste. The surnames Gosain and Maharaj declared their high status within the Hindu community. Some Brahmins were admired by the low castes, especially if they were respected older men or the sons of famous priests, and their advice on personal problems was often sought. The frequent surname or suffix 'Singh' among Kshatriyas was a mark of their superiority, but Vaishyas were relatively indifferent to caste, probably because they were not placed in an extreme social position by it. Chamars, however, were denigrated by the higher castes and sneered at when their behaviour was offensive. The very word 'Chamar' was a blasphemy and one of the worst names anyone could call a Hindu. Two important aspects of caste status are clear: caste rankings focused on the high status of Maharaj and the low status of Chamar; caste might become crucial for Hindus who aspired to community leadership.

San Fernando's castes and varnas have been divided by me into three social categories: stratum I, the upper class, comprised owners of large properties and businesses, executives and professional people who earned an annual income of more than $7500 (Trinidad & Tobago; £1500); stratum II, the middle class, was made up of white-collar workers, proprietors of small shops and properties, teachers and skilled and semi-skilled workers, with an income range between $1500 (£300) and $7500 (£1500) per annum; and stratum III represented the lower class, and comprised vendors and unskilled labourers, all of whom received less than $1500 (£300) per annum.

Out of the 149 persons whose caste, income and occupation have been compared, 9% belonged to stratum I, 36% to stratum II and 53% to stratum III (Table 5.7). The Maharaj caste accounted for almost half the membership of stratum I, and Brahmins and Kshatriyas together made up ten of the 13 members. The only other members of stratum I were two Ahirs and one Gadariya: no other Vaishya and no Sudra belonged to stratum I. Stratum II contained 46 members, and 28 were Brahmin or Kshatriya: Ahirs accounted for eight of the 17 Vaishyas, but only one Sudra, a Chamar, ranked as high as this. Stratum III represented the largest and lowest group among San Fernando's Hindus. Brahmins and Kshatriyas together formed 25% of this group, Vaishyas 44% and Sudras 14%; 17% could not – or would not – be grouped by caste and varna. It is significant, however, that 11 out of the 12 Sudras in the sample fell into stratum III. There was a strong tendency, therefore, for Brahmins and Kshatriyas to be associated with strata I and II, for Vaishyas to be associated with stratum III and to a lesser extent with stratum II, and for Sudras to be linked to stratum III.

In Chapter 4 I argued that, excepting the white population, San Fernando was only weakly segregated spatially, by race, largely because each racial segment was internally stratified by class. Equally, no neighbourhood or street in San Fernando was confined to any one caste or to any set of socially equal castes, because of caste as such. However, the fact that more than one-third of the sampled Brahmins lived within a quarter-mile radius of the southern, Todd Street *mandir* points to the importance of caste in the context of religious ritual, reinforced by high-class position and the ability to buy property in that neighbourhood.

The hierarchy of caste and class among San Fernando's Hindus was less pronounced than was the Creole system of colour–class; colour–class was central to the stratification of the dominant segment in San Fernando, whereas caste was peculiar to Hindus and East Indians of Hindu extraction – but how did caste and class relate in San Fernando? The traditional link between caste and secular occupation had been dissolved, and varna stratification had partly merged with the class structure of San Fernando. High class came to imply high caste; low class, low caste. However, although caste was an attribute of social status, it applied only to the Hindu community.

Disharmony between caste and class was most acute where Vaishyas were members of stratum I and Brahmins and Kshatriyas were members of stratum III. One-third of the Brahmins and Kshatriyas were lower class, and their low social status was readily acknowledged among Creoles. Yet these Brahmins usually received favourable treatment by their class-superiors in the Hindu community and marks of respect at religious ceremonies. The social status and rôles given to low-class Brahmins and high-class non-Brahmins clearly de-

Table 5.7 Caste and class in San Fernando (sample).

Caste and varna	Stratum I	Social groups Stratum II	Stratum III	Not placed	Total
Brahmin					
Gosain	0	2	1	—	3
Maharaj	6	16	11	—	33
subtotal	6	18	12	0	36
Kshatriya					
Chattri	4	10	8	—	22
subtotal	4	10	8	0	22
Vaishya					
Ahir	2	8	11	—	21
Baniya	0	1	2	—	3
Gadariya	1	0	0	—	1
Jaiswal	0	1	0	—	1
Kahar	0	1	2	—	3
Kewat	0	0	1	—	1
Koiri	0	1	3	—	4
Kurmi	0	1	9	—	10
Loniya	0	0	3	—	3
Madrassi	0	0	2	—	2
Mallah	0	0	1	—	1
Nau	0	0	1	—	1
Patitar	0	1	0	—	1
Teli	0	2	0	—	2
Vesh	0	1	0	—	1
subtotal	3	17	35	0	55
Sudra					
Dusad	0	0	6	—	6
Seunerine	0	0	1	—	1
Chamar	0	1	4	—	5
subtotal	0	1	11	0	12
don't know	0	8	12	—	20
won't answer	0	0	1	—	1
not placed	—	—	—	3	3
subtotal	0	8	13	3	24
Total	13	54	79	3	149

pended on the racial or religious group with which they associated and, if it was the Hindu group, on the secular or religious nature of the context.

Among Hindus in San Fernando, however, caste was a social attribute in its own right. It could be set in the balance against class when defining an individual's status. Occupation and wealth determined a Hindu's social standing in secular affairs; and caste either added to or detracted from it. Nevertheless, the conjunction of high caste and class status among Brahmins and of low caste and low class status among Chamars has reinforced the social importance of caste and varna. Many Hindus in San Fernando argued that Brahmins have proved their superiority by success in competitive business, whereas the Chamars have fallen to their appropriate level. The wealth of urban Brahmins has confirmed their traditional dominance, and lack of Brahminical status has detracted from the rank of otherwise high-class Hindus.

Débé and the Christian East Indians Caste labels were retained more completely by Hindu respondents in Débé than in San Fernando, but only 17 castes were listed by Débé Hindus compared to 25 in San Fernando, and only two – one Bhar (ploughmen and day-labourers) and one Parsi (collectors of palm oil and day-labourers) – both Sudra – were unique to Débé (Table 5.6). Maharajs were far less numerous proportionally than in San Fernando, Chamars more so. Even so, Chamar representation in Débé was low compared to the incidence recorded by Klass (1961, 61) near Chaguanas.

Earlier in this chapter it was noted that the occupational stratification at Débé was poorly developed, with unskilled and personal service work employing almost half the labour force. Circumstances were not propitious for a strong caste–class relationship. No Brahmins were in white-collar work of any kind, and only two out of four respondents (or their husbands) had skilled jobs. Kshatriyas were more numerous and more diverse occupationally: none were professionals, but two were white-collar workers, five were skilled, three semi-skilled and four either unskilled or personal service workers. Only four Vaishyas were in the lowest, non-manual category, but there were five skilled, five semi-skilled and 22 unskilled and personal service workers and four agricultural labourers. The Sudra at Débé fared much better than in San Fernando. Although there were three agricultural workers and 14 in the unskilled and personal service categories, five were semi-skilled, two were skilled, three were white-collar workers and two managers. Absence of Brahmin dominance and a weak class structure had created a more fluid situation than in San Fernando – paradoxical though that may seem in the light of Débé's isolation and traditionalism (see Smith & Jayawardena 1967, 66–8, for similar circumstances in rural Guyana).

At the other extreme from the Débé Hindu sample, more than half San Fernando's Christian East Indians claimed not to know their caste. These were the offspring of Christian parents; their links with their Hindu origins had attenuated, including their memory or knowledge of caste. Nevertheless, among those Christians who knew their caste affiliation 23 *játi* were mentioned, including seven not given by San Fernando's Hindus. One Sudra caste (Bhar) and six Vaishya (Murão, cultivators; Lohar, blacksmiths; Kayat, probably Kayasta or clerks; Lalla, unknown; Dhobi, washermen; and Bind, labourers and ploughmen) were *jati* unique to the Christians. Members of all four varnas were spread approximately equally across the major occupational groups. Significantly, out of nine Christians in the professional category, not one claimed to know his caste, which could perhaps indicate that achievement of high class status was dependent either on cutting ties with Hinduism or, more likely, on having ancestors who were non-Hindu. When white-collar workers were distinguished from others, the proportion in the former category was 59% for Brahmins, 50% for Kshatriyas, 54% for Vaishyas and 50% for Sudras. Comparable percentages for Débé were 0, 29, 10 and 17; and for San Fernando Hindus 50, 48, 19 and 0. San Fernando's varna and reinforcing class hierarchy for Hindus therefore contrasted with the Débé sample where there were weak class and varna structures, and with the Christian East Indians where the relationship between them was non-existent.

Conclusion

The major racial segments of San Fernando's society were internally stratified by occupation, education and levels of living. However, to ignore intra-segmental variation and to reduce social differences solely to class would be to omit a crucial series of variables which determined the most significant social worlds to which San Fernando's residents belonged. Moreover, the analysis has revealed that many of the samples were internally stratified by variables *other than* class – colour in the case of the Creoles, and caste for Hindus and East Indians of Hindu origin – although their association with class was very close among Creoles and Hindus.

Urban Creoles and rural Hindus were obviously very different in behaviour and attitudes. In San Fernando, East Indians and Creoles had similar co-ordinate, occupational stratifications despite differing job distributions, identical – high – aspirations for their children, and lived in similar housing areas, though East Indians preferred to own property rather than to rent. Moreover, both segments eschewed overt racism. However, these similarities must not be allowed to obscure the intense competition which had already started between Creoles and East Indians – for resources in general, and for schooling, white-collar jobs and access to the professions in particular.

The retention by urban East Indians of elements and symbols of Indian culture – languages, artefacts, food, orientation to India, a sense of an ethnic past bolstered by the cinema – made for a fundamental difference with Creoles. To a lesser degree Christian East Indians shared this heritage, but Indian identity and an awareness of East Indian difference was most acute among those East Indians who had never severed their roots in the great religions of India – Hinduism and Islam.

6 *Religions, cults and festivals*

Memberships in three of the great world beliefs – Christianity, Hinduism and Islam – subdivided the East Indians. Moreover, religion in general emerged as a major factor in San Fernando's social composition (Ch. 4), and discussion of caste required reference to Hinduism (Ch. 5).

It is not enough to label San Fernando's inhabitants as Christian, Hindu or Moslem, however. Within these religions, the theology, organisation, rituals and taboos of sectarian affiliations are profoundly consequential. This chapter explores these themes before going on to consider the influence of Christianity upon Hinduism and Islam, directly and through conversion; the rejuvenation of Hinduism and Islam in Trinidad during this century; and the problem for Hinduism of its caste-exclusive priesthood. Finally, attention is given to the great public events associated with religious festivals, to the shared spirit world of Creoles and East Indians, and to religious syncretism.

Religious affiliation

Creoles in the sample in San Fernando, as in the census, were Roman Catholic (46.0%) or Anglican (33.6%), with only 9.0% involved in revivalist Afro-Christian sects. Only one Creole was Presbyterian; none was Moslem or Hindu. *Douglas*' Christian attachments were almost identical to the Creoles', but one was a Moslem and another a Hindu. In Débé, on the other hand, Hindus (70.6%), Presbyterians (14.7%) and Moslems (10.1%) predominated among East Indians, and the handful of members of other Christian churches living in the village were not East Indian.

Hindus and Moslems, too, were subdivided, but by sect. In San Fernando, the orthodox Sanathanists were by far the largest group, and accounted for 75% of Hindus, followed by five Arya Samajis, five Seunerinis, five Kabir Panthis and one Punjabi Guru Nanic (who together made up just over 10% of the sample). Twelve per cent of Hindus claimed they did not know their sect. A similar pattern of sects obtained at Débé: Sanathanists accounted for 65.5% of Hindus, followed by ten Agor Panthis, eight Arya Samajis, eight Seunerinis and four Kabir Panthis (34.5% of Hindus *in toto*). Unorthodox sects were therefore much more prominent at Débé, where the lower castes were numerous and Brahmin-led Sanathanism was commensurately weaker. In San Fernando, however, although Brahminism strengthened orthodox Hinduism, a minority of Hindus had all but abandoned their religion.

Moslems in San Fernando were split into two major sects, each affiliated to a town-centre mosque belonging either to the more liberal Trinidad Moslem

League (TML) or to the conservative Anjuman Sumat Al Jamaat (ASJA). Approximately 40% of Moslem respondents belonged to each organisation, with 8.7% replying that they did not know their sect – presumably an indication of their lapsed status – and 6.3% declaring themselves members of the Kadiani reformist organisation; (virtually all practising Moslems eschewed pork irrespective of sect). Apart from the Kadiani, the rest were Sunni; the Shia sect did not exist in Trinidad. Similar proportions of sect members were recorded among Moslems at Débé, though their total number in the sample was only 11.

Christian East Indians, too, were subdivided – by denomination. Presbyterians formed the majority of Christian East Indians in San Fernando (65.6%) followed by other Christians (7.8%) – mostly members of the Church of the Open Bible – and Anglicans (7.0%). Christian East Indians retained membership of the original missionary denominations, though the US Protestant sect of the Open Bible had begun to make inroads into the Presbyterian congregation.

Religious affiliation has been generally stable from generation to generation, especially among Creoles, and, to a lesser degree *Douglas*, but a notable degree of intergenerational change characterised Christian East Indians, among whom 29.7% of mothers and 31.3% of fathers were Hindu. Conversion also took place on a modest scale at Débé, where the percentage of Christians in the respondent's generation (14.7%) was almost twice the proportion among their parents (8.3%). Fewer than 5% of San Fernando's Hindus and Moslems had affiliations different from their parents.

However straightforward these patterns of religious affiliation appear, religious reality was more complex. For example, one young East Indian reported he had been baptised an Anglican because he had been ill as a baby and his parents thought conversion might lead to his recovery; however, he still lived as a Hindu. Another East Indian couple described themselves as Hindu, but all ten of their children were Presbyterian or Pentecostal. Not only did converted East Indians revert, but most Hindus and Moslems – even those born in the country – went to Canadian Mission Indian Schools until the Moslem and Hindu cultural revival got under way in the early 1950s. Thus 'devout Hindus and Moslems with only a primary education [were] often quite able to discuss fairly detailed aspects of Christian theology' (Niehoff & Niehoff 1960, 151). Hinduism and Islam were influenced by the Christian ethos of Trinidad during the 19th century, yet both religions survived denigration as heathen by the white élite, the Creole masses and the Canadian missionaries (Samaroo 1982).

Retention of oriental religions and sects

Despite their trans-shipment to Trinidad, the trauma of indentured labour and loss of contact with India, East Indians retained the two great religions of their ancestral homeland – Hinduism and Islam. Hinduism was more changed than Islam by the transplantation. The monotheism of Islam, its emphasis on one sacred book – the Koran, its organisation into *jamaats*, its corporate acts of worship and its hostility to Christianity, gave its adherents a strong sense of identity and protected Moslems from the proselytising activities of the Presbyterian missionaries in San Fernando. Hinduism, by contrast, was polytheistic, diffuse, a household religion with variations by sect. Many Hindus converted to Christianity (Ch. 2) or were at least influenced by its teachings, but most Hindus ultimately adhered to the *Ramanandi Panthi* propagated since the early 1950s by the Sanathan Dharma Maha Sabha:

> Islam is a static religion. Hinduism is not organized; it has no fixed articles, no hierarchy; it is constantly renewing itself and depends on the regular emergence of teachers and holy men. In Trinidad it could only wither; but its restrictions were tenacious (Naipaul 1962, 82).

Sanathanism in San Fernando was a ritualistic, peasant sect untouched by philosophical Hinduism, though most Hindus were able to expound the three facets of God as creator (*Bramha*), preserver (*Vishnu*) and destroyer (*Siva*). The central text of Trinidadian Hinduism was not a philosophical treatise, but Tulsi Dass's epic *Ramayana*, which recounts the marriage of Rama and Sita, their adventures culminating in Sita's kidnapping by the King of Ceylon and her rescue by Rama with the aid of the warrior monkey, Hanuman. In the mid-1960s many Hindu men used to gather in the evening for *Rameyn satsang*, during which verses of the epic were read and then chanted to musical accompaniment – harmonium, *dhantal* and *tabla*; after each verse the pundit would give a translation from Hindi into English. The habitual mode of address among Hindus was '*Sita Ram*', and Rama, Sita, Lutchman (Rama's brother) and Hanuman held central positions in temple decoration and domestic rituals.

Despite high-caste pressure to conform to Sanathanism, other small Hindu sects were represented in San Fernando. Kabir Panth was founded in India by a 15th-century prophet who advocated one living God and rejected idols and priests. The Seunerini sect, created by a 19th-century Rajput called Siva-Narayan, believes in one God; their worship involved a meeting at which Siva-Narayan's sacred writings were read and chanted. The Seunerinis in San Fernando – and Débé, where they had their own temple and maintained animal sacrifice to Kali – were low caste (Sudra) and had their own *mahants* (priests), though Brahmin Sanathanist pundits usually officiated at their weddings. Arya Samaj, the most recent Hindu reform movement, was introduced into Trinidad only in the 1910s. Its Gujerati Brahmin founder espoused a return to the Vedas, the earliest Hindu scriptures. Strongly monotheistic, the sect is opposed to idolatry and caste. To my knowledge, however, no Kabir Panthi, Seunerini or Arya Samaji rituals were held in San Fernando during the period of our fieldwork.

Sanathanists in San Fernando regularly sponsored household ceremonies dedicated to a specific *deota* in propitiation or as a fulfillment of a promise (Jha 1973). The three most popular types of *puja* were dedicated to Hanuman, to *Satnaryn* (god of Truth) and to *Suruj Puran* (the Sun god) (Jha 1974). Commonly the ceremonies were performed as a sequence by one of the town's two Brahmin pundits, starting with the Hanuman *puja* (or *Rōt*) on Saturday and finishing with the Suruj Puran on Sunday. After each ceremony sacred food (*persad*) was given to participants and friends who attended, and prayer flags on long sticks, or *jhandi*, were raised in the garden or yard at the front of the house – red for Hanuman *puja*, white for the Satnaryn *puja*, white for Suruj Puran and yellow for *Lakshmi puja* (in celebration of the goddess of Light and Knowledge).

Devout Hindus marked each significant household event – including business trips off the island and secondary school scholarships won by their children – with a *puja*. Information collected in the survey showed 0.8% of Moslem respondents, 0.9% of Creoles, 2.6% of *Douglas*, 3.5% of Christian East Indians, 53.7% of San Fernando Hindus and 60.9% of Débé respondents lived in houses with *jhandi*. As *jhandi* are left to weather and decay, and cloth disintegrates quite rapidly in the tropics, it is likely that most Hindus organised one household *puja* every two years. However, to many younger Hindus, these *pujas* were without religious meaning, as Naipaul testifies, referring to his own youth: 'I barely understood the rituals and ceremonies I grew up with . . . My Hinduism was really an attachment to my family and its ways, an attachment to my own difference' (Naipaul 1983, 15).

Devout Hindus – typically Brahmins – also maintained a *puja* room or corner or a section of their yard for making daily ritual ablutions and doing *aarti* (worship) before the gods who, if the altar was indoors, were usually represented pictorially or as statues – Rama, Sita, Hanuman, Lakshmi, *Radha*, Krishna and *Ganesh* (with the elephant trunk). Many garden altars had Shiva *lingam* stones which were 'fed' daily by ritual outpourings of water from brass *lotahs*; many Hindu yards were planted with sacred tulsi trees. Some Hindus also sponsored occasional public celebrations which ran for a week and involved several officiating priests. Most popular were *Rameyn Yagya* and *Bhagwad Yagya*, the latter accompanied by the *Goberdhan puja* on the Thursday evening. These events were housed either in a specially constructed 'tent' at the sponsor's house or in the local *mandir* (temple).

In contrast to the home-based rituals of Hinduism, Islam in San Fernando was traditionally mosque-orientated, though Hindu *yagya* had their parallel in Koranic readings or *kitabs*. San Fernando Moslems subscribe to the five pillars of Islam: *kalima* (faith) – there is no God but Allah, and Mohammed is his prophet; *namaaz* (prayers); *roza* (fasting); *haj* (pilgrimage); and *zakat* (compulsory charity). Most Moslems, however, found their daily secular routine too exacting to maintain the five-prayer domestic ritual laid down for the faithful, but the devout regularly attended the Friday (sabbath) *Jumaa namaaz* at one of San Fernando's two mosques, led by the *imam*.

Both the ASJA and TML mosques were breakaways from the Takveeyatul Islamic Association (TIA) founded by Abdul Aziz from Iere Village near Princes Town in 1926. The more conservative ASJA was incorporated in 1935. After years of internal strife, Ameer Ali left the TIA in 1947 to form the more reformist TML. The TML approves of equal rights for women, includes them in the congregation and other meetings (the ASJA segregates its congregation and screens off women participants), counts eight rather than 20 *rakaats* in a prayer cycle, and believes, as the ASJA does not, that Mohammed is the one and only prophet (Smith 1963).

Conversion and Christianity

Conversion to Christianity provided all East Indians with possibilities for a secondary education and social mobility, and Hindus with an escape from caste and a more Westernised life-style including more freedom in dress and hairstyle for women.

The Canadian Mission had an enormous impact on East Indians over the century 1868–1968, and was a major factor enabling urban East Indians to compete socio-economically with Creoles on fairly equal terms in the decades preceding independence. By the mid-1960s, however, the pace of conversion had begun to slacken. Fewer than 50 converts were baptised each year at San Fernando Susamachar Church, Hindi services having been abandoned six years previously.

Converts of earlier generations not only maintained ties with their non-Christian families of origin, but retained many of the customs, tastes, attitudes and values of the Hindu community into which they had been socialised as children. They tended to favour sex segregation and social distancing between men and women, and many of the latter still wore the *oronhi* (veil). In addition, they persisted in eating Hindu-preferred foods, and a minority also avoided pork and beef.

Rejuvenation of institutions

After World War II, at the very time that traditional links with India were attenuating, local East Indian organisations, notably the Sanathan Dharma Maha Sabha and the ASJA, received renewed impetus. The stimulus was political, and operated in two ways. Adult suffrage and decolonisation encouraged the East Indian and Creole segments to organise so as to compete for political power; and, as government was increasingly managed by locals, financial help was given to Hindus and Moslems to develop institutions which taught Hinduism and Hindi, Islam and Urdu in addition to standard requirements of the secular school curriculum.

Although there was no Maha Sabha branch in San Fernando, through the organisational skills of a Chamar civil servant, Nanan, and with donations from local Brahmin businessmen, an old temple on Todd Street was renovated and re-opened in 1961. Hinduism in San Fernando rapidly became 'institutionalised', featuring corporate worship in the *mandir*. Treated like a church, the *mandir* increasingly became the centre for other Hindu activities; and the adjoining Gandhi *ashram* was used for lectures, services, occasional weddings and major events of the Hindu calendar, such as *Jhanam Astamie* (the birth of Lord Krishna).

A *saddhu* lived at the San Fernando temple, tended the gods in the *mandir*, maintained the ablution of the *lingam* stone and cleaned the *ashram*. However, as San Fernando's two practising pundits were unwilling to take regular Sunday-morning services (see also Schwartz 1967b), Bisram, leader of the Selfless Service Divine Mission based on Corinth to the east of San Fernando, was invited to hold a *Sandhya puja* each Sunday morning at 8.30. The mission, founded in the mid-1950s and with several branches in the Naparima sugar belt, carried out household and community services in San Fernando and the surrounding villages. A dozen young men and women from the mission provided music and singing, and conducted the ritual of the *hawan* (sacred fire).

Bisram, a middle-class Ahir, was the mission's inspiration and principal speaker as well as group chauffeur. He was typical of younger Hindus attracted to philosophical Hinduism through the lectures of reformist Hindu missionaries who visited Trinidad from India after World War II (Klass 1961, 149–52). In Bisram's opinion, Brahmin priests had misused their position and influence among Hindus to make money; they had not taught their followers about the meaning of Hinduism – on the contrary, it had been in their interest to keep the people ignorant and superstitious. Hinduism therefore needed rejuvenation but without abandoning Sanathanism for the Arya Samaj. He concluded that Hindus could not afford to allow energetic and earnest young non-Brahmins, like himself, to be excluded from the priesthood. 'Can't caste be achieved during a man's lifetime?' he asked. 'Why should caste depend on a man's actions in his previous incarnation?'

Bisram's *Sandhya puja*, which drew about 50 participants each Sunday morning, was reinforced by a weekly class given in the San Fernando *ashram* by an Indian headteacher from Princes Town. This class on the *Bhagavad-gita* – until then an almost unknown text among Hindus in San Fernando – had an audience of 50–100, mostly men; it met on Monday evenings, lasted about two hours, and involved a reading and interpretation of the *Gita*, with elaborate excursions into Hindu philosophy and Christian theology. Though highly abstract, this proved to be a *tour de force* that attracted Hindus and Christian East Indians, even from neighbouring villages, and began to put Hinduism on an intellectual footing previously unknown in San Fernando.

Such was the impact of the Indian teacher that the first Hindu secondary school in San Fernando, Tagore College, was built over part of the Gandhi *Ashram*, and he moved from Princes Town to become the principal. Funded by local subscription, this fee-paying enterprise boasted several hundred pupils by the later 1960s.

By the postwar period the Moslems in San Fernando already had two mosques, so the ASJA responded more quickly than Hindus to possibilities for school building. In the early 1960s a Moslem primary school was attached to the ASJA mosque, and ASJA College – a secondary grammar school – had been completed. Rumour had it that a substantial contribution made by the PNM government to the ASJA College building fund in the late 1950s had secured the Moslem vote in San Fernando at the 1961 election.

The proselytising Church of the Open Bible also merits attention as a revitalising religious force in San Fernando. Founded in Trinidad in the early 1950s by missionaries experienced in India, the Church had its headquarters in Des Moines, Iowa. Its 'Mountain Movers' service attracted Hindus and Presbyterian East Indians by emphasising healing and being born again in Christ. However, as one of the pastors remarked, 'The only way we reach a Moslem is through a miracle.' In the early 1960s about 600 persons attended the various Sunday services, about 75% of whom were East Indian.

Hindu caste, religion and the priesthood

Reform presented special difficulties for Hinduism that it did not for Christianity or Islam, because an hereditary principle – caste – was intrinsic in Hindu religious organisation. However, in religious as in secular aspects of caste the emphasis had shifted away from the twice-born (Brahmin, Kshatriya and

Vaishya) towards a distinction between Brahmins and Kshatriyas, on the one hand, and the remainder, on the other. For example, hardly anyone except male, teenage Maharajs and Kshatriyas took the *janeo* (sacred thread), and only pundits, who were expected to remain ritually clean, wore it permanently (Jha 1976a). San Fernando Hindus still believed that after death the soul returned to earth to be reborn in another body: sin in a previous life was punished by a life of suffering in the next. However, very few except a handful in the two highest varnas justified the caste hierarchy itself by past deeds.

Concepts of purity and pollution which had largely disappeared from secular Hindu life were retained in an attenuated form in situations associated with ritual. For example, priests raised their *janeo* over one ear when they had been temporarily polluted during their ablutions. The meal that accompanied the *puja* was always strictly vegetarian, and most Hindus refrained from eating flesh on the day on which they were holding a private *puja*, though otherwise they may have habitually eaten meat, including, in some cases, beef and pork. However, so intensely related was diet to pollution, that Sudras, it was said, could become Vaishyas if they gave up the habit of eating pork.

Commensality was normally practised at Hindu dinners, but some of the older, more conservative pundits insisted on being fed before, and separately from, lower-caste people. Some Brahmins also raised objections when castes lower than the twice-born, especially Chamars, or Moslems cooked the food at Hindu weddings. The view was often expressed that 'a Maharaj feels unhappy about cleanliness in the home of a Chamar'. As an extension of the link between Brahminism and purity, some Hindus believed it was their religious duty to feed and give presents to Brahmin males on the occasion of family *pujas*.

Brahmins controlled the Gandhi Service League, which ran the Todd Street *mandir*, and provided most of the money for refurbishing it in 1960–1. They accounted for at least 80% of the congregation of 40–50 at the Sunday morning *Sandhya puja*. Brahmins also held all the offices in the Stri Sevak Sabha, a women's organisation that arranged services and celebrations in the *mandir*. Indeed, most of the members of the Sabha were Brahmins or Kshatriyas. In 1964, when *Jhanam Astamie* was celebrated by a small pageant held in the *ashram*, more than 80% of the participating children were Brahmin or Kshatriya.

Brahmin control of the priesthood in San Fernando was complete. No Brahmin pundit would teach the ritual of the *pujas* to non-Brahmins, and non-Brahmins would never have been admitted to the Pundits' Council of Trinidad and Tobago, which was part of the Maha Sabha. Both of the practising Sanathanist pundits who lived in San Fernando were Brahmins, although one had a second wife who was an Ahir. Technically speaking, membership of the Pundits' Council was more crucial than being a Brahmin, because pundits could be licensed as marriage officers only with the Council's recommendation; but caste determined recruitment to the priesthood, as the following case reveals.

A *saddhu* from San Fernando, who I shall call Goberdhan, attempted to prove that his father was a Brahmin. When he was examined at a public meeting, his evidence was rejected. Thereafter Goberdhan ceased to officiate at public Santhanist rites, though he continued with his religious and magical practices at his own small *mandir* in the south of the town.

Brahmins were caught in a dilemma. They objected to non-Brahmins becoming priests, yet very few young Brahmins were being prepared for the priesthood: they wanted to get higher education and remunerative jobs in the professions, commerce or government. It was not essential for Brahmin pundits to be full-time, but all priests had to be trained. To provide a formal induction that would be attractive to such young men, one of the wealthiest Brahmins in San Fernando proposed to build a small seminary on his property to educate them in Hindu philosophy and ritual. However, this scheme immediately came under attack from non-Brahmins on the grounds that it represented reinforcement of Brahminism. Bisram, for example, argued that the theological school should be open to interested youth of all castes.

Bisram is worth considering at greater length, because he was already a victim of caste. He was indispensable to the temple Brahmins' *Sandhya puja*, though one of the Sanathanist pundits had unsuccessfully tried to whip up Brahmin hostility to him. Bisram's indispensability enabled him to penetrate the exclusive circle of San Fernando's Maharajs. In this he was helped by a close friend and associate who was a middle-class Brahmin. Bisram was a remarkable person in his own right, and a vegetarian who neither smoked nor drank alcohol. However, despite his conduct, his fluency in Hindi and English, and his knowledge of the Hindu scriptures, he would never be accorded the office of pundit. Bisram was respected as a good Hindu because he had '*sanskritised*' his behaviour and values, and lived more like a Brahmin than most Brahmins: but the final cachet of Brahaminical status was withheld since he was born a Vaishya.

Festivals

Two aspects of festivals deserve attention: the extent to which traditional East Indian and Creole events have been maintained in San Fernando; and the degree to which Creoles and East Indians participate in one another's festivals.

Carnival

Trinidad's most famous festival is carnival, a pre-Lenten celebration introduced by French Catholic planters and rapidly appropriated by the emancipated slaves (*Caribbean Quarterly* 1956, Hill 1972). Blacks converted carnival into a riotous street pageant on to which they grafted the torch parades and stick fights of their older Canboulay (*Cannes brûlées*) processions. Masking, incendiarism and mocking of the élite were vehicles for social protest and the settling of scores. After numerous attempts to modify or proscribe carnival, the authorities finally abolished Canboulay in 1884. Carnival, however, remained firmly established in lower-class Creole culture, and calypso singing became a highly prized accomplishment. At least four carnival bands, each based on a locality, existed in San Fernando in the second half of the 19th century – Jacketmen on Coffee Street, Danois on the wharf, Rose of England in St Joseph's Village, and Diamonds (whose territory is not known). This list does not exhaust the groups involved: in 1875 two other bands, Hirondelles and Jockey Boys, were involved (Brereton 1979, 167).

Carnival had escaped official persecution by the beginning of the 20th century. Organised carnival competitions were held in San Fernando as early as 1919, with prizes offered for fancy dress; original and topical bands; the best decorated motor car, horse cab and motor cycle; the best costume and best calypso; and a carnival King and Queen were selected (Ottley 1971, 128). The status of carnival, however, remained insecure until after World War II, when steel pans, made from oil drums, made their first appearance on the streets, and political changes associated with decolonisation elevated carnival into a government-sponsored national event: brown and black Creole culture became national culture (Braithwaite 1954).

My wife's diary entry describes the King and Queen competition at the Naparima Bowl in San Fernando in 1964.

> The bowl was fairly packed, most of the white and lightest people sitting in the covered seats. All the others, us included, were in the bleachers. There were hardly any East Indians there. . . . The men's costumes were particularly fine. Julius Caesar, Richard Lionheart, and the Indian Chief stick in my mind. Some of the headdresses were most finely made, wrought in brasses, gold, silver.

Carnival activities begin with *jouvert* (*jour ouvert*) and *ole mas*, which take place on the morning of carnival Monday, and conclude with Tuesday *mas* on Shrove Tuesday (Mardi Gras). Virtually everyone in the San Fernando survey, irrespective of race and religion, had watched Tuesday *mas* at some time in their lives; at Débé the figure was slightly lower. Almost as many in San Fernando had seen the more disreputable *jouvert*, which involves a fancy-dress competition in old clothes (*ole mas*) and tends to be dominated by the capers – and malevolence – of *jab molassies*, though only 42% at Débé claimed ever to have witnessed it.

However, participation in carnival in 1964 sharply differentiated the racial and religious samples. Seventy-two per cent at Débé took no part whatsoever (as spectators or participants) in *jouvert* in 1964, and the proportion declined via Hindus in San Fernando (42%), Christians (35%), Moslems (31%) and Creoles (21%) to *Douglas* (12.8%). Spectators at *jouvert*, ranged from 76% for *Douglas* to 24% for *Débé* residents. Percentages for those participating as costume wearers and steel bandsmen followed the same sequence as for spectators; not a single respondent, whether in San Fernando or Débé, participated in a steel band. Evidence for the peak celebration of Tuesday *mas* produced exactly the same pattern of sample ranking for those who did not take part, but the urban samples all scored high turn-out rates; even Débé residents, in their majority, had watched the carnival parade in San Fernando. Steel-band membership showed identical scores and ranking to those for *jouvert*, and costume participation in Tuesday *mas* was only marginally different.

Over 90% of Creoles and *Douglas* agreed that carnival 'was the greatest show on earth', but only 35.8% at Débé. A similar ranking appeared in reply to the question of whether East Indians should take part in carnival: the range was from 82.1% in agreement among *Douglas* to only 38.3% for Hindus in San Fernando. A mere 9% of Hindus in San Fernando agreed that carnival was too dangerous, compared to 58.7% at Débé; but only 6% of Creoles opined that carnival should be left to them, compared to 60.6% at Débé.

Creoles were willing to include other groups in carnival, but Débé Hindus were self-excluding. In nearly all aspects of carnival, Hindus in Débé and San Fernando expressed rejection and non-participation. By contrast, *Douglas* and Creoles viewed carnival as a central expression of their culture and were deeply involved as participants to varying degrees. Urban Moslems and Christian East Indians inclined toward partial acceptance and involvement. The extent of urban East Indian participation in Tuesday *mas*, both as spectators and as members of costume bands, however, ran counter to some of these attitudinal responses. East Indians disapproved of the month-long bacchanal of carnival, but usually participated in carnival Tuesday.

With regard to steel bands and calypso, Creoles showed absolutely no desire for cultural exclusivity: Creole culture in their eyes was national culture, and open to all. Urban East Indian opposition to this point of view was muted, and even rural Hindus expressed considerable ambivalence, since calypsos on East Indian topics frequently appealed to them.

All Saints Night

The only other Creole religious festival of note which involves substantial ritual – except Christmas and Easter, both of which are secular celebrations for Hindus and Moslems – is All Saints Night. This Catholic ceremony of grave decoration, usually with lighted candles, commemorates the dead, and is said to involve the participation of all racial groups. More than half the Creole respondents took part in All Saints Night in the November preceding the survey, and just under half the *Douglas* and Christian East Indians. For rural and urban Hindus and Moslems, however, the participation rates were much lower – 21.1%, 18.1% and 15.1% respectively. There was a distinct split between Christians and non-Christians, which is repeated for Hindu and Moslem festivals.

Diwali

Undoubtedly the most popular Hindu festival in Trinidad, as in India, is *Diwali*, or the festival of lights (Jha 1976b). Levels of participation in the 1963 festival were 91% for San Fernando Hindus and 79% for Débé residents, 11% for Christian East Indians, 7% for *Douglas*, 5% for Moslems and zero for Creoles. Marginal participation among non-Hindu East Indians occurred because of invitations to house celebrations offered by Hindu kin and neighbours. San Fernando, because of its high castes, was more culturally conservative than the rural areas when it came to orthodox religious behaviour.

Other Hindu festivals

Both *Siw Ratri* and *Ram Naumi* (celebrations of the births of Shiva and Rama, respectively) were moderately supported by Hindus – a little more so in San Fernando than in Débé – and participation rates of about 1 in 3 were recorded, dropping to less than 1% among Moslems and Christian East Indians and to zero for *Douglas* and Creoles. This pattern was repeated for *Ramlila*, the Rama pageant, but with lower rates for urban and rural Hindus. One black respondent objected that *Ramlila* was overtly racist and ought to be banned by the government. 'Ramlila is a heathen celebration. The main purpose of this celebration is to teach their children to hate Negroes. The large idol which they make [Rawan] they claim to be a Negro God which is burned afterwards.' Creole turnout for *Kartik Nahan*, the festival of the sea, was also nil, whereas urban Hindu participation soared to 58.4%. *Holi*, *Phagwa*, or 'Indian carnival' as it is often called, hardly existed in San Fernando but flourished in Débé where 50% of respondents were involved – as *abir* (purple dye) stains on shirts and dresses confirmed for many months afterwards.

For all these Hindu festivals, Creole and *Dougla* participation, as revealed by the survey, was zero. Orthodox Sanathanist festivals, such as *Kartik*, were strongly maintained in Brahmin-dominated San Fernando, but less so in rural Débé. Conversely, the more rustic, secular *Holi* was widely supported in Débé.

Moslem events

The major celebration of the Moslem year, in San Fernando as elsewhere, is the month-long fast of *Ramadan*, during which orthodox Moslems must eschew food and drink during daylight hours. Just over half the San Fernando and Débé Moslems had submitted to this test of faith in the year before the survey, but less than 1% of Christian East Indians or Hindus was involved and no *Douglas* or Creoles. *Eid ul Fitr*, the feast that ends the fast, was more widely supported, over 94% of urban and rural Moslems marking the celebration, with a handful of Christian East Indians, Hindus and *Douglas* sharing in the festivities. *Buckra Eid*, the celebration of the story of Abraham and Isaac, received fractionally weaker support from all parties, though it recorded exactly the same pattern of scores as for *Eid ul Fitr*.

All these festivals are Sunni events the world over. A Shia festival, *Hosay*, commemorating the martyrdom of Hossain, Mohammed's grandson in AD 680, which marks the beginning of the great schism in Islam, was presumably introduced into Trinidad by a handful of Shia Moslems, and has been incorporated into Sunni folkways as a largely secular spectacle. The major *Hosay* event takes place at St James, in Port of Spain, with East Indian, *Dougla* and Creole participation, but there is considerable support for it in the rural areas, too, including Débé. In San Fernando, 10.3% of *Douglas*, 6.7% of Hindus, 6.3% of Moslems, 1.6% of Christian East Indians and 0.9% of Creoles took part. *Douglas* had a tendency to attend festivals irrespective of the religion being celebrated, but Moslems in the case of *Hosay* were ambivalent. Orthodox Sunni businessmen placed adverts in leading newspapers congratulating the Moslem community on the event (for commercial reasons, of course), only to reveal in private their distaste for this heresy.

Hosay reflects cultural transfer. Rejected by the staunchest of urban Sunni, it has been maintained by urban *Douglas* and creolised East Indians, and by rural Moslems who have often followed a heterodox theology to the dismay of visiting *mullahs*. *Tadjahs* – wood, paper and tinsel representations of the tomb of the martyr – were paraded through the streets (but not in San Fernando) and were eventually thrown into the sea or rivers. Drinking, drumming and stick fighting traditionally accompanied the festival (Khan 1961, Naipaul 1976), and Creole youths have often taken part in *tadjah* construction as well as in the more secular activities.

East Indian festivals have remained largely incomprehensible to Creoles, and it took the need for a vote-catching gesture in the mid-1960s to lift *Diwali* and *Buckra Eid* to the status of national holidays. Moreover, unlike carnival, the East Indian festivals have only occasionally been expressions of social protest. For example, the *Hosay* riot in San Fernando in 1884 was precipitated by a government decision to prevent rural Moslems from carrying their *tadjahs* into the town, though 'one of the objects of the 1884 regulations was to prevent Creoles from participating in [Hosay], and gradually after 1885 they withdrew from it' (Brereton 1979, 184).

Evidence from religious festivals supports, for the East Indians, Crowley's observation that 'all members of any group know something of the other groups, and many members are as proficient in the cultural activities of other groups as of their own' (Crowley 1957, 819). However, few Creoles were involved in Hindu or Moslem public festivals, and the rôle of San Fernando Creoles even in the ambiguous *Hosay* celebration was negligible (Glazier, 1983b).

Spirit world

One area of religious activity in which Crowley's (1957) comments carry weight centred on belief in the spirit world. East Indians who believed in the magical properties of Hanuman had little difficulty in absorbing black Creole ideas about *jumbies* (spirits), *lagahoos* (werewolves) and *soucouyants* (vampires) (Herskovits & Herskovits 1947). East Indians, fearful that their babies might be harmed by an evil eye (*meljeu*), protected them by putting a string of jet beads on their wrist. Exorcism of spirits, particularly the *patna* that made East Indian girls barren, was the task of the *ojha* man, but manipulation of the spirit world, in malevolent as well as protective ways, was thought to be more within the competence of *obeah* men – usually Creole blacks (Vertovec 1984), engaged by both East Indians and lower class Creoles to secure their promotion at work, success in examinations, recovery from illness and triumph in love affairs.

Hindus had a choice of three remedies for illness – a *Durga puja* prepared by a Sanathanist priest, a visit to an *obeah* man or an *ojah* man, or a consultancy with a qualified medical practitioner; sometimes individuals would try all three. For minor ailments *jharay* (sweeping away) was considered effective by Hindus, Moslems and Christian East Indians, and mothers, grandmothers and *saddhus* were thought to be able to cure simple complaints, such as headaches, by holding the temple of the sufferer in one hand and saying a prayer (often backed up with an aspirin).

Syncretism

A much more important issue than attendance at one another's ceremonies is culture change and acculturation in the main bodies of religious belief and practice. Culture change among immigrant groups may be divided into persistence and loss, and persistence can be further categorised as retention, reinforcement and syncretism, or synthesis, of the ancestral culture with that of the 'host' society (Smith 1965a, 24–5). East Indian religions in San Fernando express substantial, though modified, retention; reinforcement – and alteration – by visiting Indian 'missionaries'; and, as we shall see, some syncretism.

If syncretism is thought of as putting old wine into new bottles, the Hindu *puja* provides a fine literal example. Rare was the ceremony during which oblations of *ghee* (clarified butter) were not poured on to the sacred fire from a half-pint rum bottle, which had become as indispensable a receptacle to pundits as a brass *lotah* or *tarriah*. Again, Christian mortuary practice had influenced Hindu and Moslem burial, and Paradise Cemetery served as a last resting place for all sections of the community. Traditional Hindu cremation was revived only in postwar years, using the Creek to the south of San Fernando as the site for funeral pyres (Plate 2).

Syncretism also occurred between the pantheon of Hindu gods and Catholic saints, though this was apparent only to Hindus. Hindus frequently equated Christ to Krishna, and Hanuman was syncretised with St Michael, the warrior, and *Ogun*, the *Shango* god of War, beloved of black Creole cultists (Simpson 1964). Perhaps the most outstanding example of syncretism in the San Fernando area is *Siparu Mai*. The patron of the Catholic Church of La Divina Pastora in Siparia, located about 20 km south of San Fernando, this black madonna is revered by Creoles and Venezuelans as well as by Hindus, who identify her with the Hindu goddess, Kali (who is also black).

A Hindu couple I met at La Divina Pastora explained the significance of Siparu Mai:

> People came to pray to Siparu Mai to help them out of trouble, to give them health, and especially to give them children. Children who were the gift of Siparu Mai were shaved outside Church.

The festival of *Siparu Mai* occurs on Maundy Thursday and Good Friday, and is distinct from the Creole fiesta of *La Divina Pastora*, which is held two weeks later (Niehoff & Niehoff 1960, 155). By Maundy Thursday evening, Siparu Mai has been removed from her niche in the church and placed on a platform in the wooden school room in the yard to the rear. When I arrived at 8.45 a.m. on Good Friday, I noted in the diary that

> crowds of people were filing through, making money gifts to Siparu Mai (collected by her custodian, an elderly Negro lady) and offering sweet oil and candles . . . Money was usually handed to the woman, but some coins were thrown to Siparu Mai from the body of the school room. Only half the sweet oil was poured out for Siparu Mai. The remainder, which is regarded as holy oil, was returned to the doner for his own medicinal use . . . After making their offerings to the madonna, and saying prayers, the devotees moved slowly down the line of beggars (some of whom had come from the sidewalks of San Fernando), giving 2 cents and a handful of rice to each person. Some of the mendicants wore *dhoti* and *kurta*, and one or two had on turbans and *janeos*.
>
> We went back outside and looked for the Naos (barbers). We found them in the far corner of the school yard. Most were giving male children a short back-and-sides . . . One little boy who had hair down to his shoulders (indicating this was his first hair cut) was completely shaved (Plate 3). His mother took the hair and carried it to the end of the school room where

Siparu Mai stood (Jha 1976a, 44). She laid it in the dust under the building and covered it up. Little Creole boys waited until her back was turned and then scuffed the hair up. But to their dismay, there was no sign of the traditional silver coin offered by parents.

There were now several hundred people in the sandy school yard. A cluster of about 50 had gathered around two East Indian male dancers dressed as women.

One dancer was quite stout. He wore an *oronhi*, shirt and long skirt. His cheeks were rouged and he had gold fillings in his teeth. The other dancer was slim and rather effeminate, wearing a false bust and anklets. Unlike his companion, his *oronhi* was topped by a crown . . . He wore a blue bodice and a long, flowing pink skirt. While he danced quite elaborately with hips swaying sexily to the music, the other man would take up children and dance with them. Their mother or grandmother put some money in a handkerchief spread as a receptacle, and then bent down and spread her *oronhi* on the ground. The man danced on the *oronhi*, holding the child and singing to the music provided by a hand organ and drums. At the end of the dance the man bent down and pressed his thumb into the dust and made a mark on the baby's forehead and on its mother's.

Devotees of Siparu Mai were of both sexes and all ages, but almost all were East Indian – as our survey data were later to confirm. Eleven per cent of San Fernando's Hindus, 9.2% of Débé residents and 1.6% of Moslems participated, yet none of the other samples reported their attendance. Siparu Mai, though a syncretic cult, has been incorporated into lower class Hindu folkways, leaving La Divina Pastora to Creole Catholics.

Conclusion

Religious affiliation reinforced segmentation of East Indian from Creole. Despite Presbyterian missionary proselytising, the two great oriental religions introduced by indentured Indians – Hinduism and Islam – persisted: indeed, missionary visitors from India (and later Pakistan) and local reformist leaders (and reactionaries) had strengthened them. Conversion to Christianity had slowed to a trickle by the mid-1960s, but Presbyterians formed the largest East Indian segment in San Fernando and the most Westernised and creolised. Islam had been less disrupted in the past by conversion than Hinduism; and Hinduism was experiencing tensions generated by its caste-exclusive priesthood. Some involvements across race and religion were recorded, especially of East Indians in the Creole festivals of carnival and All Saints Night. However, Creole participation in Hindu and Moslem festivals was negligible, even in *Hosay* and *Siparu Mai*. Asymmetry in this respect reflected Creole control of the national society; East Indians were expected to comply. Typical of the incomprehensibility of East Indian culture was Creoles' inability to distinguish between mosques and temples in San Fernando, and their ignorance of which religion each represented. The lowest-class East Indians and Creoles found common ground only in their belief in the supernatural; however, the convergence of these two groups is found also in some aspects of household composition.

7 *Household, kinship and marriage*

Segmentation of Creoles and East Indians is mirrored by household structure, kinship and marriage (Matthews 1953, Niehoff & Niehoff 1960, Freilich 1961, Klass 1961, Braithwaite & Roberts 1967, Roberts & Braithwaite 1967a, Rodman 1971). The voluminous literature on Caribbean Creole households has focused on the black lower-class family, whose major features have been described as female household headship, visiting unions, serial polygamy, late marriage, the prevalence of half-siblingship, and the tendency for individuals to move from extra-residential mating to consensual cohabitation and marriage (Freilich 1961, Roberts & Braithwaite 1967b, c, Rodman 1971). By contrast, upper- and middle-class Creole marriage and mating approximate to European norms, though the taking of black or coloured mistresses is quite common. Caribbean East Indian household composition, though morphologically similar to the European, has more complex mechanisms for social control (Smith and Jayawardena 1959). Early arranged marriages, patrilocal households, male headship, the absorption of the elderly into their children's homes, and a more marginal rôle for common-law unions are the major features (Roberts & Braithwaite 1962). How far do circumstances in San Fernando fit these generalisations?

Marital condition and parental status

Households were defined in this study as consisting of people who sleep in the same shelter, eat from the same pot, and, when able, contribute to the unit's domestic costs. According to the survey data, the largest households on average were recorded at Débé (7.1 persons) followed by urban Moslems (6.4), *Douglas* (6.4), Hindus (6.1), Christian East Indians (6.0) and Creoles (4.9). Proximity of Christian East Indians to the Creoles and polarisation between Creoles and Débé residents repeated patterns recorded in previous chapters. Two sets of data have been analysed for each household sample (see p. 84): the first relates to household heads, and explores their parental status and marital condition by age and sex (Appendix C, Table C.1); the second makes the same comparisons for all adults in the household other than the head (Table C.2).

Just over 72% of Creole household heads were men – the lowest proportion recorded in the various samples (MacDonald & MacDonald 1973). Among these male-headed units, the most common status category was married person with resident spouse (66.5%) (Table C.1). Marriage was the norm for young and mature adult males, and the rate increased slightly with age. Disintegration of marital and consensual unions was quite common, however. Respondents who had been married but were separated and not single (6.9% of heads) were more common than those in their first common-law union (6.3%); the proportions of the once married but separated and single, and of those who had cohabited consensually and were separated but not single were both 3.2%.

Creoles were outstanding for the frequency of female household headship (21.2%). Among these Creole female heads, 45.8% were widows, 18.6% had been married but were separated and single, 10.1% were single persons without mates whose parental status was not known, 8.5% were single parents but no longer mating, and 6.8% were separated from common-law husbands and bringing up children on their own. In short, female headship was strictly associated with households from which men were absent through death or breakdown of union, and with the existence of visiting relationships. As for male heads, there was no striking temporal sequence from extra-residential unions via consensual cohabitation to marriage such as that recorded for other Caribbean Creole communities.

Women, of course, predominated among dependents in Creole households (Table C.2): most were spouses (37%), followed by young single women who were not mating (22.2%) or whose mating status was unknown (17.7%). Male dependents, on the other hand, were mostly concentrated in the age category under 24: 42.4% were not mating; 36.1% were of unknown mating status; 7.0% were in a mating relationship. As Creole men entered their mid-twenties, they departed from their parents' households to head their own domestic units. Wherever Creole men were co-resident with their mates, they were normally household heads. Exactly half the women aged 24–39 were married, and this proportion rose to 65.7% in the age group 40–54. Young spinsters concealed the offspring of their mating by placing their children with kin: thus whereas three women aged under 24 were single parents, the number increased to only eight for those aged between 24 and 39.

The *Dougla* sample is too small to warrant close analysis. The outstanding features were the high proportion of women dependents who were married and living with their spouses (69.6%), and the large proportion of men who were in their first common-law unions (17.4%).

East Indians can be treated as a single group and between-sample differences merely highlighted (Table C.1). About 80% of East Indian male heads were married and cohabiting, and only among urban Hindus was the percentage as low as 72.1. East Indian male heads involved in their first common-law union were barely half the Creole proportion (6.3%) except at Débé (6.0%), and only among San Fe nando Hindus did the percentage of heads who had separated from concubir es and were cohabiting anew (2.2%) approach the Creole rate. Single parents and single persons who were mating were rare among Creole heads, and non-existent among the East Indian samples.

Not only was female headship unusual among East Indians, but, Moslems excepted, it never occurred where a male spouse was resident. Thus although widows comprised the largest element among Creole female heads (42.4%), widows exceeded 60% for all the East Indian samples except the Christians (51.2%). The proportion of female heads who were married, separated and single was lower for Hindus, Moslems and the Débé sample than for Creoles, Christian East Indians and *Douglas*, and single parentage occurred half as frequently among Christians and Moslems as among Creoles; at Débé and for Hindus the rate was zero.

Turning to dependents in the various household samples, Christian East Indian men were outnumbered by women in every age group (Table C.2). Nevertheless, compared to Creole men, Christian East Indians were outstanding for the large proportion of male adults who were not mating and for the substantial number aged below 40 who were living as dependents, even when married to a resident spouse. The mode for Christian East Indian women was to be married (46.6%) and the next most common category was the group that was single and not mating (30.4%). Over 68% of women aged 24–39 were married (18% higher than for Creoles), but there were also large numbers of celibate male and female household dependents aged over 25. Only 0.5% of East Indian Christian women were single parents, compared to a rate of 4.8% among Creoles.

Male dependents were more numerous than females in the Hindu group aged up to 24, because of the early marriage age of girls: for all other ages women outnumbered men. Hindus fitted more closely the Christian East Indian than the Creole pattern, with a high proportion of young men not mating and some once-married men living as dependents of household heads. Among Hindu women, marriage was even more common than for Christians, and single parenthood was rare (0.8%). Indeed, 65.2% of women aged 25–39 were married, the rate rising to 72.2% among women aged 40 to 54.

Females outnumbered males in all age groups among Moslem dependents, and the pattern was almost identical to that for Hindus, except that married women and male dependents were more prominent (51.6% and 20.9%, respectively) and single parents and consensual couples even rarer. Débé dependents resembled those in the urban East Indian groups, but higher marriage rates for male and female dependents were recorded: celibacy and virginity among the young were prescribed. Yet there is also evidence for the greater social acceptability of consensual cohabitation among East Indians in Débé than in San Fernando: 5.2% of women dependents in Débé were in their first common-law union; indeed, 1.2% (both were women aged under 24) were single parents. However, consensual cohabitation was not a prelude to marriage for young rural Hindus: seven out of nine women in common-law unions were aged over 25 and five of them were over 40.

Differences between Creoles and East Indians in marriage and parenthood characteristics merit restatement. Marriage in San Fernando was more strongly associated with East Indians than Creoles, among whom female headship was common but by no means the norm (Bell 1970). Moreover, East Indian households were on average larger than their Creole counterparts. The single-person household was a markedly Creole feature, but East Indians frequently engaged in concubinage, and it would be wrong to think of even serial polygamy as confined to Creoles. At Débé, serial polygamy was negligible compared to its incidence among Creoles in San Fernando; but urban Hindus were outstanding for the close association of concubinage with the breakdown of a previous marriage – a condition in which their experience directly contrasted with that of Débé respondents.

How did class structure affect consensual cohabitation? Concubinage was a lower-class phenomenon at Débé as well as among *Douglas*, Hindus, Moslems and Christian East Indians. However, almost 60% of Creole heads who were in

Plate 4 Wedding pole in *maro* (wedding booth), Hindu wedding. Note the mortar, pestle and rum bottle.

Plate 5 The *milap* (meeting up) of the bride and groom's entourages, Brahmin wedding, San Fernando. Many of the men are wearing the *dhoti* (loincloth).

Plate 6 *Bedi* (altar) for use at the *dwar puja* (ceremony of the gateway), Hindu marriage ceremony. The brass vessel is a *lotah*.

Plate 7 Groom wearing *maur* (crown) and visor, at *dwar puja* (ceremony of the gateway), Hindu wedding.

consensual unions were skilled workers or had higher ranking, and among them concubinage was culturally rather than economically determined.

The marital and parental status of adults is more similar in all the urban samples than the literature on East Indians and Creoles suggests. This is due largely to the economic affluence of San Fernando, and to the financial ability of low-class Creoles to afford early marriage. However, institutional practices that are much more salient in other low-status Caribbean Creole communities – concubinage, serial polygamy, female headship and extra-residential mating – do occur among San Fernando Creoles but remain marginal to mainstream East Indian practice. East Indians emphasise early marriage and male headship, and avoid extra-residential mating and female headship, except in the case of widows. Young married East Indian couples are kept under the watchful eye of the groom's parents, and many neolocal households are not created until the husband is aged over 40.

Relationship to the household head

Creole male-headed units had three sorts of dependents – mates (20% of dependents), lineal issue (60%) and mates' issue (9%) (Table C.3). Brothers and sisters of the heads and their offspring were rarely dependents. Lineal issue (72%) dominated the dependents among female-headed units, but siblings (5%) and their children, mothers and matrilateral kin, together with unrelated persons, were numerous (16%).

Other noteworthy categories of dependents in male-headed Creole domestic units were mates' sons by others (2.0%), mates' daughters by others (2.2%) and mates' daughters' children (1.1%). All other categories were miniscule. In female-headed units (27.2% of the Creole total) the most prominent categories of dependents were sisters (4.1%), sisters' children (3.6%), brothers' children (2.6%), and mothers (2.6%).

There were only two cases of the father of a Creole household head being present – one for males and one for females, but female-headed households had greater generational depth: mothers (2.6%), daughters' sons (3.6%), daughters' daughters (7.1%), daughters' grandchildren (1.0%) and sons' daughters (1.0%) were persistent categories, yet were insignificant in male-headed units.

These San Fernando household data involve units drawn from across the range of the Creole stratification, from upper to lower class, from white to black (Ch. 5). To assess upper- and middle-class norms in the sample data a brief comparison is made with M. G. Smith's (1962, 163–99) sample of black lower-class households in Kingston, Jamaica. In Kingston, female household headship was more common than in San Fernando, accounting for almost 50% of units; spouses and mates were more prominent and lineal issue less prominent than in San Fernando's Creole households; the category 'other non-kin' was identical for male heads and similar for female heads in both towns. San Fernando, as previously suggested, approximated more closely than Kingston the European norm of the patrifocal, nuclear, biological family irrespective of class, yet matrifocality and the housing of non-kin remained characteristic of some of its black Creole domestic units.

How did *Dougla* and East Indian households compare with the Creoles? Among *Douglas*, spouses, mates and lineal issue accounted for over 90% of dependents in male-headed units, and mates' sons and daughters made up a mere 2.8% – figures very much at variance with the Creole data. The paucity of female heads (though similar to the proportion among East Indians) must be borne in mind when analysing these aberrant statistics on dependents. Lineal issue were far fewer than in any other sample, including the Creole, and sisters, brothers, brothers' children, sons' mates and spouses, and sons' mates' children by another together accounted for 35.6% of dependents.

About 90% of dependents of Christian East Indian male heads were spouses and mates or lineal issue – the biological family *par excellence*. Among other dependents, brothers (1.1%), and daughters' mates and spouses (1.0%) deserve mention, though there was a wide range of more distant kin.

Moslem heads' dependents, too, were predominantly spouses, mates and lineal issue (87.8%). Noteworthy additional members were mates' sons by others (1.7%), daughters' mates and spouses (1.3%) and mates' sisters and brothers (1.2%). As with Christian East Indians, female heads recorded almost four times the proportion of grandchildren as were listed for males.

Hindu male heads were similar to Moslems in the proportion of their dependents accounted for by spouses, mates and lineal issue (87.6%). The only other categories of significance were sons' mates and spouses (2.8%) and sisters' children (1.1%). Female-headed units had one and a half times as many grandchildren as male-headed homes, a ratio higher than for Moslems and Christians. Other important categories of dependents were sisters (6.1%), sisters' children (3.0%), mothers (3.0%) and sons' mates and spouses (3.0%) – a similar pattern of kin as for Creoles and Christians. However, non-kin and adopted persons were far fewer in all urban East Indian units than in Creole homes.

The urban Hindu and the Débé samples, as noted previously, recorded the lowest proportion of female household heads; unlike all the others, in neither of these two cases was there a single female head with a resident spouse or mate, and headship of homes was unequivocally a male prerogative. At Débé 85% of male-headed domestic units comprised spouses and mates and lineal issue of the head; other categories of importance were sons' mates or spouses (2.6%) and mates' sisters' children (2.1%). Lineal issue accounted for 85.1% of the dependents of Débé's female heads, and grandchildren were five times as frequent, proportionately, as in male-headed homes. All other categories were very small.

A number of generalisations can be made on the basis of these data. Female headship was much more common among Creoles than East Indians; indeed it was particularly rare among rural East Indians. A major distinguishing feature of male-headed households among Creoles was the dominant position occupied by their mates' issue and kin – an indication of serial polygamy. Among East Indians, not only was male headship the norm, but Hindu female headship entailed the complete absence of men unless they were descendants; in contrast, patrifocality ensured that only the man's lineal issue were present. However, although female headship among Hindus and Moslems occurred with a frequency less than half that for Creoles, irrespective of race or religion there was everywhere an emphasis on lineal issue, materterine and uterine kin and on the placement of grandchildren. Non-kin were common dependents among Creoles and Débé villagers, but rare in the urban East Indian samples. All East Indian units, irrespective of religious affiliation or whether they were male or female headed, included large proportions of sons' mates and spouses, thus confirming that it was rare for East Indian marriages to lead immediately to the creation of neolocal households.

Types of domestic unit

Domestic units in San Fernando have been divided into a dozen categories ranging from single person units to couples and issue to the fourth generation – plus other, more complex, types (Table 7.1). Couples with children were the mode for all segments, though the percentage was nearly twice as large for *Douglas* (71.4%) as for Creoles (39.6%), with the East Indian segments fitting in between. Indeed, there was a minor difference between the Creoles and urban East Indians in this respect (Hindus, 52.3%; Christians, 55.3%; and Moslems, 61.4%); and between Creole and urban East Indian segments on the one hand and Débé villagers on the other. In Débé, three- and four-generational units together accounted for more than 30% of household types, compared with 14.7% for Creoles and 20.6% for San Fernando Hindus.

Table 7.1 Differing types of domestic units in the San Fernando and Débé samples.

Types of domestic unit	Samples (%) Creole	*Dougla*	Christian East Indian	Hindu	Moslem	Débé
single persons	10.6	3.6	3.7	5.8	0.0	0.9
siblings	1.4	0.0	3.3	2.6	1.5	0.9
household head and children	17.1	14.3	9.0	6.5	5.3	7.3
childless couples	12.0	3.6	7.8	9.7	7.6	5.5
couples and children	39.6	71.4	55.3	52.3	61.4	49.5
household head and grandchildren	0.0	0.0	0.8	1.9	0.0	0.9
couples and grandchildren	1.8	0.0	0.4	1.9	0.0	0.0
household head, children and grandchildren	3.7	3.6	5.3	3.9	6.1	9.9
couples, children and grandchildren	4.6	0.0	6.1	11.6	11.4	12.8
household head and issue to fourth generation	0.5	0.0	0.0	0.6	0.8	0.0
couples and issue to fourth generation	4.1	3.6	4.5	2.6	3.8	10.1
other	4.6	0.0	3.7	0.6	2.3	2.8
Total	100.0	100.0	100.0	100.0	100.0	100.0

The incidence of households based on siblings, on household head and grandchildren, on couples and their grandchildren, and on household head and issue to the fourth generation was very similar among the samples. However, Creoles were noteworthy for the number of childless couples, and Creoles and Christian East Indians were distinguished from the other segments by the complex 'other' category. Creoles recorded almost twice as many single-person households (10.6%) as the next group (Hindus 5.8%), whereas the score for Moslems was zero. Single-generation households were strongly associated with Creoles (19.8%), though Hindus (14.2%) and Moslems (11.5%) recorded moderate percentages. The mode for all segments was two generations, and three generations occurred twice as frequently at Débé (36.7%) as among San Fernando Creoles (18.0%), yet no differences in this respect between Creoles and urban East Indians were detected. The truncation of Creole units was neatly expressed by the high incidence of units consisting of household heads and their children (17.1%) a proportion almost double the Christian Indian score (9.0%), and more than double those of other urban and rural East Indians, irrespective of religion.

Placement of children

Couples with children formed such a small proportion of domestic types among Creoles that only 44.2% of the children of Creole heads were living with both parents. Among East Indians the proportion ranged from 61.9% for Hindus to 74.2 for Moslems. In every sample except among Débé residents, children were placed with mothers rather than with fathers, and this was especially noticeable among lower-class Creoles. Creole fathers remained unimportant for residential child placement even when mothers were no longer alive.

The principles involved in child placement in the various segments can be studied further by analysing the distribution of dependents aged under 24 years who were separated from both parents (Table 7.2).

Creoles were characterised by the high proportion of non-kin foster-parents with whom children separated from both parents were living. At Débé, on the other hand, mother's sisters or parents were the focus of child placement, thus enhancing the generational spread in the receiving households. Urban East Indians followed the Creole pattern of non-kin adoption; the rôle of the mother's sister and parents in the placement of kin was less important than it was in Débé, and much more emphasis was given to the child's own sister and its mother's brother. Data on patrilateral placements revealed not only fewer children than for matrilateral ones – half as many for Moslems and Creoles, one-third for Hindus and the Débé sample – but the operation of different principles. Creoles placed children with their father's mothers and to a lesser degree with their father's parents or sisters, yet at Débé all three of the possibilities were completely avoided in preference for placement with agnatic or cognatic kinsmen. Urban East Indians followed a modified version of the Creole system – Christian East Indians being particularly creolised in this respect – but Hindus and Moslems were, in addition, remarkable for the placement of children with their brothers and father's brothers.

Table 7.2 Placement of children less than 24 years separated from both parents in the San Fernando and Débé samples.

Guardians	Samples					
	Dougla No. (%)	Creole No. (%)	Christian East Indian No. (%)	Moslem No. (%)	Hindu No. (%)	Débé No. (%)
Matrilateral kin						
mother's mother		10 (17.2)	2 (7.7)	4 (22.2)	7 (17.9)	1 (5.3)
mother's mother and father		4 (6.9)	4 (15.4)	3 (16.7)	4 (10.3)	7 (36.8)
mother's sister	1 (33.3)	8 (13.8)	3 (11.5)	3 (16.7)	5 (12.8)	8 (42.1)
mother's father		2 (3.4)	1 (3.8)		5 (12.8)	1 (5.3)
matrilateral kinswoman		7 (12.1)				1 (5.3)
cognate kinswoman		1 (1.7)				
non-kin adopted children	2 (66.6)	16 (27.6)	7 (26.9)	5 (27.8)	5 (12.8)	
own sister		1 (1.7)	5 (19.2)	1 (5.6)	6 (15.4)	
mother's brother		1 (1.7)	2 (7.6)	1 (5.6)	5 (12.8)	
mother's mother's mother and father			1 (3.8)			
mother's brother's father-in-law			1 (3.8)			
mother's mother's sister					2 (5.1)	
mother's mother's brother						
wards and others		8 (13.8)		1 (5.6)		1 (5.3)
Total	3 (100.0)	58 (100.0)	26 (100.0)	18 (100.0)	39 (100.0)	19 (100.0)
Patrilateral kin						
father's mother		7	2		2	
father's mother and father		2	2	1	1	
father's sister	1	3	4	4		
agnatic kinsman						4
cognatic kinsman			3			2
father's brother		1		3	4	
brother			2	1	4	
father's mother's brother				1		
Total	1	13	13	10	11	6

Union status

Creole and East Indian households in San Fernando were similar because of the widespread emphasis on early marriage and the nuclear biological family, but among low-class Creoles the high incidence of female household headship, of non-kin in households, of extra-residential mating and marriage breakdown produced differences between the two major segments in size and type of domestic unit. Minor variations also occurred among the urban East Indian samples and between them and the Débé sample. The high degree of similarity in general between Creole and East Indian household types in San Fernando was due not to acculturation, but to the existence of European family norms in the Creole upper class; to the financial capacity of lower-class blacks to emulate them – especially with respect to early marriage; and to the absence of the three- or four-generation extended family – so typical of Débé – among urban East Indians. So far, however, nothing has been said about marriage, and to this key institution we now turn.

With the notable exception of the Hindu (44.5%) and Débé (40.4%) samples, between 70 and 80% of household heads were legally married by religious rites. Because of the comparatively recent enactment of Hindu and Moslem marriage ordinances, customary – religious – but illegal unions were recorded by 9.8% of Moslem heads, 21.3% of Hindu heads and 37.6% of Débé heads. Fewer than 15% of heads in all samples had been lawfully wed without religious rites, the figure dropping from 14.6% for urban Hindus to 0.9% for Creoles. Unions not sanctioned by law were commonest among Hindus (16.8%), the Débé sample (15.6%), Creoles (15.2%) and *Douglas* (14.3%) and rarest among Moslems (8.3%) and Christian East Indians (7.8%). At Débé and among urban Hindus, common-law unions were associated with poverty. However, whereas concubinage at Débé was regarded as a socially acceptable substitute for marriage, for Hindus in San Fernando, it usually followed a marital rift. Consensual cohabitation was an ingrained Creole cultural trait, and, among the lower class an acceptable alternative to matrimony that might be a prelude or sequel to marriage.

It was less in the fact of being wedded or not than in the *method* of courtship and marriage that Creoles and East Indians differed most markedly. From these differences stem many others that have been analysed above. Thus the Creole ideal marriage age for men was 26–30 years and for women 21–5 years: each was five years above the East Indian – and *Dougla* – norm. Attitudes to dating and courting varied accordingly. Creole and *Dougla* parents allowed dating, though some respondents specified that it should take place in the presence of a chaperone; but over 40% of Hindus and Débé residents opposed all forms of dating, with Moslems and Christian East Indians only slightly more tolerant. Creoles and *Douglas* thought it perfectly acceptable for young people to choose their marriage partners, and Moslems and Christian East Indians allowed their children to find their own spouses subject to parental veto. However, most Hindus and Débé villagers abided by the Indian custom of parental choice, modified by the child's opinion. Parental control over the choice of marriage partner among Hindus in particular and East Indians in general helps to explain the early age of marriage, the prevalence of marital unions over consensual cohabitation, and the patrilocal residential basis of the family. It has also reinforced entrenched patterns of racial avoidance on the part of Hindus, Moslems and even of Christian East Indians when contemplating marriage partners.

Marriage among East Indians

Courtship and marriage among San Fernando Creoles followed familiar West European norms which require little description; attention is focussed here on the more unusual institutions of the East Indians. Child marriage died out after World War II, but at the time that the fieldwork was carried out it was not unusual to meet middle-aged – and even younger – women who had been married off by their parents at 11 or 12 years. One woman of 60 had been betrothed at seven.

A 30-year-old Hindu doctor, who had just returned to San Fernando from specialist medical training in Britain, contemplating marriage to a well-educated Christian spinster, expected it would be arranged by his father. Both potential partners were of high-class Brahmin origin. In a similar vein, a Hindu friend in his early twenties told me that he hoped to get married soon to 'a good girl, fair skinned, without secondary education, but able to cook, sew and look after a home': he had contacts, particularly his brother-in-law at Tunapuna, looking for such a girl for him.

More typically, the Hindu girl's father would look for a boy to whom to marry her. The search began when she was about 18; the father used his kin and friendship networks to seek out potential husbands, taking into account their similarity to his daughter in race, religion and class, the preference for village exogamy or town exogamy in San Fernando and, among Hindus, the couple's compatibility according to their birth-dates, checked on the pundit's *patra* (horoscope) (Jha 1976a, 46). Once a suitable boy was located, his father would go to a neighbour's house in the girl's village and ask discreetly about her and her family. If the replies were satisfactory, he would go on to the girl's home, looking carefully at the maintenance of the yard and buildings, inside and out. The two fathers would meet, and an excuse would be made to call the girl into the room for a few moments without her realising what was happening. When she had come and gone, the fathers would talk over the dowry and begin to conclude their financial arrangements.

Within a week or two the boy would be brought to the girl's home, giving them an opportunity to talk briefly on their own and make up their minds whether to proceed with the marriage. By the mid-1960s it was unusual for boys and girls never to have met before the wedding ceremony itself, and most parents were willing to give them a right of veto over the marriage: indeed, many arranged marriages involved parental validation of love relationships that had already been established (Davids 1964, Nevadomsky 1981a). However, it was quite common for educated girls to put off arranged marriages by pleading the need for further professional training or by pointing out that a clutch of GCE O levels had prepared them for a more sophisticated life-style than their intended husband could sustain.

Once the *patra* had indicated a suitable date for the wedding – usually a Sunday in the sugar-harvesting season, when money was more plentiful – the ceremony of orthodox, Sanathanist Hindus, irrespective of caste, followed a set pattern (Smith & Jayawardena 1958). Shortly before the wedding a *tilak* or bethrothal was held at the home of the bride's parents. Preparations for the wedding ceremony itself began soon after, with *matti kore* or planting of the nuptial pole and preparation of parched rice or *lawa* for use in the marriage ritual. On the day before the wedding a *katha* or reading from one of the sacred Hindu texts took place at the bride and groom's homes, and on the wedding eve, at the bride's parents', the men prepared a feast (their traditional responsibility), cooking enormous quantities of vegetarian food – *parata roti*, rice and curry.

The wedding took place in a booth, or *maro*, beside or behind the bride's parents' home. The *maro* usually had a decorated bamboo frame and a central, nuptual pole sheathed in banana stems interlaced with flowers. At the foot of the marriage pole an altar, or *bedi*, decorated with dyed rice, was constructed (Plate 4). The wedding ceremony itself began on the arrival of the groom's entourage (traditionally called a *barat*) (Jha 1974), with Indian film music blaring from loudspeakers mounted on the leading car. Outside the bride's home, the *milap* took place (Plate 5): a meeting up of the couple's fathers, each accompanied by kinsmen, friends, pundits and furiously playing drummers. The bride's mother and her friends then tempted the groom and his *siballa* (best man, traditionally his younger brother) out of their car by giving them presents. The groom eventually emerged garbed in a long pink or yellow gown, with an elaborate crown or *maur* and a veil over his face.

At the entry to the bride's yard the officiating priest performed the *dwar puja* or ceremony of the gateway (Plate 6). This act of welcome involved a gift of money to the groom and the washing of the big toe of his right foot by the bride's father and her eldest brother (Plate 7). The bride was then seen by participants for the first time. She entered the *maro* wearing a yellow sari and sat between her mother's legs; the *Nau's* wife symbolically cut their toenails and painted red dye on their feet. The groom's elder brother garlanded the bride (*dal puja*) and presented her with 'going-away clothes'.

The central part of the ceremony opened with the *kanya dan*: the bride sat on her father's left thigh, symbolically referring to her birth, while the groom perched on the wedding bench (*pirah*), made without joints from a single piece of wood. This 'giving away' was followed by *gupta dan*, or gift in secret, of coins enclosed in a ball of dough, the latter being shaped at the top like the lips of a vagina. The bride then sat on the *pirah* on the right hand of the groom in preparation for the *pau puja*. During this the bride's father washed the groom's big toe on his right foot and presented gifts to the couple. Other relatives and friends entered the *maro* in turn to offer presents – traditionally a brass *lotah* and *tarriah*.

When all these gifts had been presented, the bride and groom followed the pundit through the *hawan* or fire ritual. The *siballa* mixed *lawa*, or parched rice, brought from each home and ceremonially burned it. Vows were exchanged in Sanskrit, translated by the pundit into English, then the groom's sash was tied to the bride's sari and the couple circled the sacred fire seven times in a clockwise direction, throwing *lawa* on the fire each time. The bride led for the first four turns, the groom for the last three. The couple were covered with a white sheet, while the bridegroom rubbed red powder (*sindoor*) into the parting of the bride's hair to signify her entry into the married state. Traditionally, the lifting of the bride's veil at this moment was the first occasion on which a groom saw his wife's face.

The bride and groom retired to the house where, in the *kohabar*, the groom was teased by the bride's female relatives and the bride changed into her 'going-away clothes' – sometimes a white wedding dress. Later the bridegroom returned with his kinsmen and friends to the *maro*, where they refused to eat the specially prepared food (*kichree*) until they were satisfied with the gifts of money made to them by the bride's father and his relatives. The groom's party prepared to depart with the bride, but before they left the groom's father entered the *maro* and shook the marriage pole as a sign of his satisfaction (*maro hilai*).

After additional ceremonies at the groom's parents' the same evening and the next day, the bride returned to her father's house for the remainder of her first week of marriage until her husband collected her on the 'second Sunday' to set up home and consummate the union under her father-in-law's roof. If the couple was well-to-do, a Western-style honeymoon of up to two weeks might take place before the bride was returned to her parents.

Popular among Sanathanist Hindus who wanted to marry by religious rites but could not afford the elaborate paraphernalia of the orthodox wedding was the *jai mal* union. In these rites a pundit would meet the couple at the bride's house and perform a marriage *puja* during which they would garland one another before signing the marriage register. At the other extreme, educated young Hindus opted for marriage in San Fernando's *mandir* – a close approximation to a church wedding – followed by a Western-style reception at a local club.

Arya Samaji weddings followed a simple ritual, not for reasons of poverty or modernity, but to exclude superstition. The marriage ceremony consisted of four parts: the *agwani* or meeting up during which the fathers garland one another; the *kanya dan* or virgin giving; the *hawan* or fire ceremony; and the *saptapadi* (seven steps) that seal the union.

The prelude to Moslem weddings was similar to that of Hindu marriages in that a prayer reading (*kitab*) was held at the bride and groom's homes. During the marriage eve, the men at the bride's house prepared the feast and cooked rice, curry, chicken, goat and *parata roti*, the latter being heated over an enormous hot plate or *comal*. The groom's entourage arrived by car the next afternoon when film music was played, but the meeting up of the bride and groom's fathers involved only the normal Islamic salutations. The groom wore a white turban and shoes, but otherwise was dressed in a dark lounge suit offsetting his bride's white wedding dress.

The marriage ceremony itself was conducted in Urdu, usually by a *hafiz*. Modern couples often sat together in the sitting room, but strict Moslems still married by proxy, with the girl in the house and the boy outside, and oral communication of the vows by a go-between. The ceremony involved the placing of a wedding ring on the bride's finger and the cutting of a multi-tiered, iced wedding cake, before the bride was driven off to her parents-in-law's house. As in the Hindu wedding, the celebration was teetotal, but alcoholic drinks were usually served to male guests at a neighbour's house.

Presbyterian East Indian marriages followed the West European norm, with several 'Indian' adaptations. As we have seen, young Christian couples exercised much more freedom of choice than Hindus in the selection of partners; moreover, the sequence of events was similar to that of other Christian courtships and weddings. However, unlike most Protestant engagements in Europe, the giving of the ring involved a prayer ceremony in the girl's home, the blessing of the ring, the cutting of a cake by the fiancés, the removal of the knife by a married couple, followed by a meal and drinks (including alcohol).

Prior to the Presbyterian wedding a tent of bamboo and galvanised iron was erected outside the girl's home – as in the Hindu case – to house the reception.

The bride and groom were dressed in entirely Western clothes. After an orthodox Christian wedding, the married pair were driven back to the bride's home in a *barat*. The reception included a variety of food – meat and rice, sandwiches, ice cream, Chinese food, soft drinks and alcohol. The cake was cut, speeches were made, and toasts drunk before the bride and groom left on honeymoon.

East Indian mixed marriages present problems. Hindus and Moslems who marry out of religion must be married in a registry office, and only Christian East Indians who marry Creoles have the option of a religious ceremony. Only one mixed marriage was witnessed in San Fernando, and that involved a Presbyterian East Indian boy who had fallen in love with a 'Portuguese–Creole' girl. The boy's father was Hindu–Presbyterian, having oscillated in his religious affiliation to advance his career as a teacher. However, the bride, who was light-skinned, was Roman Catholic, and the central core of the wedding celebration took place in her church and followed standard rites. Attached to this Christian event were a series of Hindu celebrations that the groom's father insisted on holding at his own home, including the obscene songs of Hindu *matti kore*. But none of these events had the least meaning for the groom.

Thus, however similar Creoles and East Indians appeared to be in many aspects of their household composition, wedding rites involved racial exclusivities among the two segments. Among Hindus, however, marriage was not simply a question of keeping family and kin connections within the racial and religious category: it also involved issues of caste and varna. So when Hindu fathers in San Fernando looked for grooms for their daughters, they were motivated to select young men of identical caste or similar caste status to themselves. However, despite such parental control, matching brides and grooms by caste proved extremely difficult, especially in the Vaishya varna, where so many castes had small numbers of members. Thus selection of spouses from the same varna, rather than caste (*jat*) became a social rule of thumb among Hindus.

Caste and varna endogamy and exogamy

The caste of 96 men and 96 women who were engaged in marital or consensual unions has been set out in Table 7.3. Out of these 96 Hindu couples, exactly half had married caste endogamously and half exogamously. The frequency and caste range of exogamy differed very greatly among castes and between men and women of the same caste. The experience of the men is examined in detail.

Of the 24 Maharaj men, 15 married Maharaj women, four married Chattris and five married Vaishyas. No Maharaj had taken a Chamar bride. Of the three Gosain men, two married out of caste, but to Maharaj women in both instances. Of the 15 Chattri men, 12 married endogamously and three exogamously, two to Vaishyas and one to a Chamar. Caste exogamy was particularly high among the castes of the Vaishya varna, 31 men out of the 43 having married out of caste. Eight of the thirteen Ahir men, six of the nine Kurmis and two of the three Baniyas married exogamously, as did all the men of the Gadariya, Kahar, Kewat, Mallah, Lunia, Teli, Nau and Madrassi castes. Although village (and town)

Table 7.3 Caste and marriage in San Fernando (sample).

Caste of Hindu women	Caste of Hindu men																		
	Gosain	Maharaj	Chattri	Ahir	Baniya	Gadariya	Kahar	Kewat	Koiri	Kurmi	Lunia	Madrassi	Mallah	Nau	Teli	Vesh	Dusad	Chamar	Total
Gosain	1																		1
Maharaj	2	15										1			1				19
Chattri		4	12	2			2			1	1								22
Ahir		1		5			1			2	1	1			1				12
Baniya				1	1					1									3
Barhai						1				1									2
Gadariya		1																	1
Kahar		1																	1
Koiri									2										2
Kori														1					1
Kurmi		1	1	1	1			1	1	3									9
Lunia																	1		1
Mali										1									1
Nau		1							1										2
Sonar			1																1
Vesh															1				1
Dusad				2													3		5
Chamar			1	1	1								1				2	5	11
?Dass				1															1
Total	3	24	15	13	3	1	3	1	4	9	2	2	1	1	3	0	6	5	96

exogamy was usually practised in Trinidad, the small size of the last eight castes and the absence of strong caste feeling in the Vaishya varna probably accounts for this high rate of caste exogamy. Although three of the six Dusads married endogamously, the highest record for caste endogamy was found among the Chamar men. Not one of the five had taken a non-Chamar spouse. The nearest approach to orthodox caste endogamy occurred in the Maharaj, Chattri and Chamar castes.

It has already been demonstrated (Ch. 5) that varna closely approximated class in San Fernando. In fact, the ranking of castes in the varna system was more important for most Hindus than caste as such. This is clearly shown by the fact that although 47 of the 96 Hindu unions that have been analysed were caste endogamous, 70 were varna endogamous and only 26 varna exogamous. Of the 27 Brahmin men, nine married out of varna: four married Kshatriyas and five married Vaishyas. Not a single Brahmin married a Sudra woman. Of the 15 Kshatriyas, 12 married endogamously and only three exogamously, two to Vaishyas and one to a Sudra. No Kshatriya man married a Brahmin woman. Although caste exogamy was high among the Vaishyas, varna exogamy was not. Only 13 of the 43 Vaishyas married out of varna, six to Kshatriyas, two to Brahmins and five to Sudras, again illustrating the median position of the varna. Out of the 11 Sudra males, ten married endogamously and only one exogamously, to a Vaishya. Of the Brahmins, 33% married out of varna, as did 20% of the Kshatriyas, 30% of the Vaishyas and 9% of the Sudras. The fact that varna exogamy occurred more than twice as frequently between adjacent varnas (eighteen examples) as between non-adjacent varnas (eight examples) provides further evidence for the hierarchical ordering of the varnas.

The four varnas can be used to measure the relative frequency of *hypergamy* and *hypogamy* among Hindus in San Fernando. Almost twice as many women as men married into varnas higher than their own. Furthermore, the caste range of hypergamous unions was greater than for hypogamous ones. No Kshatriya man married into the Brahmin varna, but four Kshatriya women did. Two Vaishya men married Brahmins, but five women did. No Sudra man married a Kshatriya, but one woman did. One Sudra man married a Vaishya, but five women did. The only varna that did not accept a Sudra woman in a marital union was the Brahmin. With the exception that six Vaishya men and only two Vaishya women married into the Kshatriya varna, women of lower varnas were able to enter unions with higher-varna mates far more frequently and more easily than men. There were two cases of hypergamy within both the Brahmin and Sudra varnas to confirm this point.

Hypergamy plays such a large part in varna exogamy because Hindu children inherit caste patrilineally. No person in the sample adopted his or her mother's caste, although some Hindus may have attempted to adopt their mother's higher and more prestigious caste. Caste pride and feelings of superiority made Brahmin fathers reluctant to 'down-caste' their daughter's children, and lower their own standing, by arranging hypogamous marriages. However, unions with lower-caste girls, especially if they were light-skinned, beautiful, wealthy, or had a secondary-school education, were much less objectionable. Outstanding personal qualities and high social class also modified Brahmin attitudes towards boys of lower castes. A pundit in the south of Trinidad married his daughter to a Chattri whom he favoured because the boy was well-placed financially and had a fair skin. Nevertheless, a husband's high-caste surname can better conceal a wife's lower caste origin, and this undoubtedly facilitated hypergamy among the Brahmins and Kshatriyas.

The high proportion of varna endogamy in San Fernando (73%) is partly to be explained by the larger size of the high varnas in San Fernando, the strength of their conservative views, and the inability of the Sudra to overcome prejudice against them. However, above all, varna endogamy has been reinforced by the high correlation between varna and class. Only one of the 26 males who married out of varna belonged to social stratum I, and he was a Brahmin. Five males belonging to social stratum II married out of varna; three were Brahmins and two were Vaishyas. However, 20 of the 26 males who married out of varna belonged to stratum III and five were Brahmins, three Kshatriyas, eleven Vaishyas and one was a Sudra. Low-class status was frequently the common denominator of varna exogamy, and high-class status was an important basis for endogamy in the two highest varnas. No low-class Brahmin boy married even a wealthy low-caste girl, though caste may have been traded for class in this way.

Varna-exogamous unions tended to be associated with conditions of cultural breakdown, particularly among the lower class. Only 19% of the unions were common-law; however, one-third of these involved varna exogamy, most of them hypergamous.

Débé and the Christian East Indians

Although almost three-quarters of San Fernando's Hindus married within varna, the figure dropped to 42.6% at Débé, a proportion that is consistent with results from another rural East Indian study in Trinidad (Schwartz 1964c, 1967a). At Débé, only Kshatriyas maintained substantial varna endogamy (80%); Brahmin matches with other varnas, particularly with Kshatriyas, were especially frequent. Exogamy was common among lower-class villagers, and women married men of lower caste than themselves (18 cases) more frequently than men married women of a lower caste (11 cases). Caste breakdown may be attributed to lack of the Brahmin domination that was so evident in San Fernando.

Christian East Indians in San Fernando experienced a weakened varna-endogamy system. Fifty-two per cent of Christian East Indian unions were varna endogamous, the rate being lowest for Vaishyas (24%) and highest for Brahmins (75%). Varna exogamy for male Brahmins and Kshatriyas grouped *together* was 19% among San Fernando Hindus, 26% for Débé and 44% for Christians. Varna (and caste) among Christian East Indians appeared to be of no significance, yet their high rate of varna endogamy reflected residual caste consciousness. No Brahmin or Kshatriya Christian man married a Sudra Christian woman (or so they claimed); the rates for San Fernando and Débé Hindus were 2.4% and 15.8%, respectively.

Conclusion

In terms of household, kinship and marriage, Creoles and East Indians in San Fernando were clearly parts of larger Caribbean-wide populations introduced into the region and subsequently moulded by slavery and indenture. A major distinction in San Fernando involved the early age at marriage and reduction of female headship among Creole blacks, associated with urban affluence. Urban East Indians were unusual in having only a restricted range of generations in their households compared to villagers at Débé (see also Schwartz 1964a), though even in rural Trinidad patriarchs rarely lived with sons and their families in joint compounds (Lowenthal 1972, 153; Angrosino 1976). East Indians at Débé were identical in household composition, marriage ritual and varna endogamy to their counterparts in rural Guyana (Smith & Jayawardena 1958, 1959 & 1967, 75).

Household data reviewed above reflect the same sociocultural continuum revealed in previous chapters. Creole and Débé household structures were most dissimilar; *Douglas* were most like Creoles and Christians in household composition; and urban Hindus and Moslems stood between the Christian and Débé samples. Segmentation depended upon the cultural matrix in which the East Indian households were set, and in particular, upon parental control of the choice of marriage partner and the persistence of distinctive wedding customs among Hindus and Moslems. In the case of Hindus, the operation of the family extended into the realm of caste and varna, and concern for caste as a pedigree was stronger in San Fernando than in Débé. Yet even among Moslem and Christian East Indians, for whom notions of caste were non-existent or minimal, early marriage guaranteed racial endogamy and the persistence of Creole–East Indian segmentation.

8 *Intersegmental association and political affiliation*

Social interaction in San Fernando takes place with varying degrees of intimacy. Where Creoles and East Indians live is determined largely by their capacity to purchase or rent accommodation: but clubbing, friendship and inter-marriage involve interpersonal interactions of greater and greater intimacy that are socially and culturally proscribed as urban residential segregation is not. In these social contexts racial and cultural divisions have encouraged intrasegmental political organisation and racial voting in San Fernando as they have in Trinidad generally. Moreover, the politics of culture and race have in turn increased social distance between East Indians and Creoles.

Inter-marriage

Most people marry their social equals: endogamy and exogamy express social proximity and social distance in the most intimate of interactions. Information for the San Fernando and Débé household samples has been analysed in three ways: by measuring endogamy and exogamy among household heads, using race and religion as the criteria; by examining the race and religion of dependents of the household head, where the head and spouse were of the same race and same religion; and by determining the race and religion of dependents where the household head had no spouse present. The measures of exogamy were made with reference to three racial segments (Creole, East Indian and *Dougla*) and three religious segments (Christian, Hindu and Moslem). Thus a Creole head who married a Christian East Indian formed a union that was racially exogamous but from the point of view of religion was endogamous. The term 'marriage', as noted in Chapter 7, presents certain problems in Trinidad. As used in this chapter it includes consensual cohabitation or concubinage as well as legal and customary marriage.

Racial and religious exogamy

The evidence relating to inter-marriage between persons of different races indicates the small scale of exogamy, even when the upper limits of the sample errors for each group are taken into account (Table 8.1). Racial exogamy was particularly uncommon for Hindus (3.1%) and for Moslems (1.8%), owing to antipathy towards blacks and to parental influence over the choice of marriage partner. Interracial marriage was conspicuously absent among Hindus in Débé, where racial endogamy was the rule.

Creoles recorded the highest rates of exogamy (11.0%), but were closely followed by the Christian East Indians (10.7%), among whom control over children's behaviour was weak or non-existent. Only among the *Douglas* did

Table 8.1 The household mix in the San Fernando and Débé samples.

Household head	Race of household head: Creole	Hindu	Moslem	Christian East Indian	*Dougla*	Débé Hindu
with resident spouse of different race*						
total households	143	129	113	187	24	72
spouse of different race – no.	17	4	2	20	14	0
– %	11.0	3.1	1.8	10.7	58.3	0.0
±2 S.E.%	5.3	2.8	2.4	4.4	0.3	—
with resident spouse of different religion†						
total households	143	129	113	187	24	72
spouse of different religion – no.	1	18	13	15	1	2
– %	0.7	14.0	11.5	8.0	4.2	2.8
±2 S.E.%	1.4	6.1	6.0	3.9	8.1	3.9
with resident spouse of same race‡						
total households	126	125	111	167	10	72
dependent of different race – no.	3	1	3	5	3	2
– %	2.4	0.8	2.7	3.0	30.0	2.8
±2 S.E.%	2.8	1.8	3.0	2.4	9.2	3.9
with resident spouse of same religion§						
total households	142	111	100	172	23	70
dependent of different religion – no.	0	19	10	11	0	5
– %	0.0	17.1	10.0	6.4	0.0	7.1
±2 S.E.%	—	7.0	6.3	4.8	—	5.2
with no spouse present¶						
total households	74	25	19	57	4	8
dependent of different race – no.	10	2	3	5	2	0
– %	13.5	8.0	15.8	8.8	50.0	0.0
±2 S.E.%	7.7	10.7	16.6	7.4	50.0	—
with no spouse present						
total households	74	25	19	57	4	8
dependent of different religion – no.	0	1	9	5	0	1
– %	0.0	4.0	47.4	8.8	0.0	12.5
±2 S.E.%	—	7.7	22.9	7.4	—	23.4

*Christian East Indian–Creole: $\chi^2=0.026$, d.f. = 1, $P>0.40$.
†Total: $\chi^2=22.18$, d.f. = 5, $P>0.001$; Hindu–Christian East Indian: $\chi^2=2.27$, d.f. = 1, $P>0.05$.
‡Total: $\chi^2=26.42$, d.f. = 5, $P<0.001$; Hindu–Muslim: $\chi^2=0.39$, d.f. = 1, $P>0.25$.
§Hindu–Muslim–Christian East Indian: $\chi^2=6.92$, d.f. = 2, $P<0.05$.
¶Creole–Hindu–Muslim–Christian East Indian: $\chi^2=4.20$, d.f. = 3, $P>0.70$.

The table contains data for independent samples. Chi-square has been used to test the significance of the difference between three or more samples and between selected pairs of samples. Where three or more samples are examined together, the average rate of, say, exogamy is calculated and used to compute the expected frequency; the Null hypotheses that are tested state that the conditions being examined are distributed at random among the samples. The tests carried out on selected pairs of samples employ 2×2 contingency tables, and the Null hypotheses argue that there is no significant difference between the values recorded in the two populations. This set of calculations incorporates corrections for continuity.

exogamy exceed endogamy, non-*Dougla* mates of *Dougla* household heads comprising five East Indians and nine Negroes.

The racial origins of the exogamous mates of Creoles and East Indians throw further light on the nature of inter-group relations. Among the 17 non-Creole spouses of Creole heads of household, 11 were East Indian and six *Dougla*. The majority of these unions were legal; consensual cohabitations were outnumbered by ten to seven. The four cases of racially exogamous Hindu heads – all male – included one white, one *Dougla* and two coloured; all these unions were legal. Both the racially exogamous unions entered into by Moslem heads were with *Douglas*, one being legal, the other not. Four of the spouses of Christian East Indian male household heads were *Douglas*, and the remaining 16 had Creole mates. Seventeen of these unions were legal. Racial exogamy in San Fernando seems to have been associated with marriage rather than with concubinage; inter-marriage was infrequent, but seldom clandestine. Only among Creoles was exogamy associated – though weakly – with consensual unions. However, this was to be expected, since concubinage occurred more frequently among Creoles than East Indians. Endogamy was the norm, perpetuating segmentation of the urban community along racial and religious lines, with Afro-Indian mixtures being shed into the *Dougla* category. *Douglas* figured prominently in the racially exogamous unions of all other groups, an indication of their broker status and small numbers.

Unions contracted across religious lines were just as infrequent, though the racial and religious samples differed significantly in this respect. In San Fernando religious exogamy was highest among Hindus (14.0%), Moslems (11.5%) and Christian East Indians (8.0%), and lowest for *Douglas* (4.2%) and Creoles (0.7%). Most of these marriages were legal. Although Creoles and *Douglas* more often married out of race than out of religion, the converse was true among Hindus and Moslems. East Indian Christians again occupied an intermediate position in the system, though with slightly higher rates of racial than of religious exogamy. However, in Débé, exogamy, limited though it was (2.8%), was expressed solely in religious terms.

Whenever Creoles married East Indians they almost invariably united with Christians: no Creole head of household had a Hindu spouse and only one spouse was Moslem. Christian East Indian males also were more than twice as likely to marry Hindus as Moslems, and this expressed the close links between Hindus and Christians – forged through conversion. Children of mixed marriages in San Fernando were more likely to adopt their mother's religion if she was Christian than if she was Hindu or Moslem.

Race and religion of dependents

When the household head and spouse were of the same race the dependents in the sample were generally identical to them. Only small numbers were racially different from the head, and the difference between the groups (as measured by chi-square) was due principally to the *Dougla* households. In none of the other groups did more than 3% of the households contain persons of a race different from that of the head and spouse: *Douglas* were prominent among these dependents. Hindu households were markedly homogeneous with regard to race, though not significantly different from those with Moslem heads. Only one Hindu household contained a non-Indian child; although he was an adopted son of Negro and white parentage, he had been brought up as an orthodox Hindu. None of the three Creole households contained non-Creole dependents related by blood to the head or the head's spouse.

Variation in the frequency of household heads having dependents of another religion was much greater. Creoles and *Douglas* did not enter into the analysis, and the rate increased from Christian East Indians (6.4%) to Moslems (10.0%) and Hindus (17.1%), with the Débé sample mean (7.1%) falling between the Christian East Indians and Moslems. These differences between East Indian samples in San Fernando were statistically significant, and most of the dependents involved were in-laws of the household head. Among the Hindus, however, 16 of the 19 cases concerned children who had been converted to Christianity.

Where household heads had no resident spouse, the percentage of households with dependents of another race was greater than otherwise for all groups except the Hindus at Débé. The rate increased from Hindus in Débé (0%) to *Douglas* (50%), but the sample errors are large, and no statistically valid distinction can be made among Creoles, Hindus, Moslems and Christian East Indians. Households probably increased in heterogeneity under conditions of family breakdown; alternatively interracial unions may have been more brittle than endogamous ones.

Presence or absence of the household head's spouse had a less clearcut impact on the religious affiliation of dependents. For Moslems, Christian East Indians and Débé Hindus, the frequency of households having dependents of a different religion increased when the head had no spouse, which was the opposite of what happened among urban Hindus. Moreover, Creoles and *Douglas* recorded a zero frequency, as they also did wherever spouses were present.

All nine Moslem-headed households with dependents of different religion from the head owed their heterogeneity to conversion or to complex family structures. Among Christian East Indians, however, most of the dependents were Hindu kinsmen. Although conversion contributed to the heterogeneity of households, the breakdown of the family, as measured by the absence of a spouse, facilitated the formation of more complex domestic units (see Ch. 7). This process is particularly discernible among the strongly paternalistic groups, especially the Moslems.

Interpretation of the data

Creoles and East Indians – Hindu, Moslem and Christian – were all endogamous with respect to the racial and religious criteria that have been applied. Although East Indians, and especially Hindus, often claimed that Creoles were bent on assimilating them through inter-marriage (Klass 1960, 859, Singh 1965, 14), little evidence supports this view. 'East Indians and Negroes should be encouraged to inter-marry so as to produce a beautiful race', claimed a young black; yet Creole heads who united with East Indians usually chose Christians, and engaged no more frequently in marriage with other races than their Christian East Indian counterparts. Indeed, Creole male heads were less likely to cohabit with Hindu and Moslem spouses than Hindu and Moslem male heads with Creole spouses, and this undoubtedly reflects the protected position of non-Christian Indian women. Hindu and Moslem heads were more likely to marry out of religion than out of race; but Christian East Indians were more likely to marry out of race than out of religion and so, too, were the *Douglas*.

Christian East Indians and *Douglas* formed links between Creoles and Hindus and Moslems, as their exogamous records suggest, and Hindus were closer to the Creoles and Christians than were the Moslems. Inter-marriage between Moslems and Creoles was virtually non-existent. Hindus in Débé represented an ultra-conservative East Indian population: racial endogamy was complete and religious endogamy was marked. This rural sample provides a contrast to the greater exogamy of Hindus in particular, and of all East Indians in general, in San Fernando.

The patterns of association revealed by household analysis are similar to those established for census data in 1960 using indices of dissimilarity, yet each method indicates differences of degree. Whereas the spatial data, underlain as they are by economic factors, show the marked absence of racial and religious segregation, except for whites, the household samples, excluding only the *Douglas*, are substantially homogeneous.

Above all, the data depict different degrees of interracial and interreligious association at the spatial level in the public domain and at the domestic level in the private domain. The comparative prosperity of Trinidad in general, and San Fernando in particular, was largely responsible for the high degree of spatial association, whereas the cultural values of the East Indian, and especially Hindu and Moslem, households were reflected in endogamy and social separation. It may be argued that the relatively high degree of residential mixing among the various races and religions is more important, and that the choice of marriage partner is a private concern. However, the Eastern religions, through racial and religious preference and parental control of the choice of marriage partner, are able to nullify the potential influence towards exogamy of spatial proximity; thus endogamy reinforces the segmentation of the community.

Friendship and clubbing

Friendship involves less intimate – and less permanent – links than marriage, for interpersonal alignments based on free association can be changed many times during a life. However, friendship in San Fernando, though less exclusive than marriage, tended to coincide with and reinforce endogamy. Forty-seven lists of friendships were collected from a subsample of the original survey respondents – 13 from Creoles, 33 from East Indians and one from a *Dougla*: each respondent listed five or six friends and provided details about their race or religion.

Friendship

Among Moslems one in every ten friends was a Creole; the ratio for Hindus was 1 : 7 and for Christian East Indians 1 : 3. Creoles reported even greater racial isolation: barely one in 15 of their friends was East Indian. The sole *Dougla* respondent, typically, had friends in both segments. However, other interpretations can be put on the results. Out of 13 Creole respondents, only two named

East Indian friends. Yet three out of six Moslems, three out of nine Hindus and four out of 18 Christian East Indians listed at least one Creole intimate. Creoles were numerous enough and politically secure enough to isolate themselves if they so wished: East Indians could not avoid contact with Creoles; indeed, some found it agreeable or useful to cultivate Creole friendships.

Preference for friends reveals well-established patterns of association. Hindus and Moslems emphasised in-group relationships, but also interacted to a lesser extent with Christian East Indians. Creoles were isolated from Hindus and Moslems, and their limited contact with East Indians was through the Christian East Indians, who once more acted as a broker element. Most Christian East Indians selected friends from their own segment, yet their Creole contacts outnumbered links with Hindus and Moslems taken together, by almost 2 : 1, which repeated, in exaggerated form, Christian East Indian preference for Creole spouses over Hindu and Moslem spouses.

Interracial friendships were rare in San Fernando, and where they occurred, class and shade configurations were usually at work, as the following case study confirms. During fieldwork we were able to observe one Creole clique in which the focal figure was a black woman whose family had enjoyed élite status since the 1920s (see Ch. 3). Among her 18 Creole friends none was darker than her, and eight were white or 'pass-as-white'. Those who were employed held high-ranking positions in private business, the legal profession and the police. The circle of mutual friends also included the Chinese widow of a well-known doctor, a *Dougla* civil servant and his East Indian, Presbyterian wife. The clique, complete or in part, met at weekends and occasionally in the evenings, each person or couple taking it in turn to hold a drinks party, which sometimes also involved a buffet, barbecue and dancing. At one party, harsh words were voiced about East Indians, and the speaker immediately added that their East Indian friend, the *Dougla*'s wife, who was present but out of earshot, was Creole in behaviour and in no way to be associated with his criticism.

Dominant members of this clique were the wealthiest and best connected. The members were all children of 'established' families, and they excluded the socially mobile, younger generation of Creoles promoted by party politics. However, the group, in turn, was kept at a distance by the handful of French-Creole whites, on grounds of their social inferiority and multiracial inclusiveness; sole social contact with East Indians was via the *Dougla* and limited to his Presbyterian wife.

Club membership

Although friendship was usually home-orientated, clubbing involved semi-public encounters. However, club entrance fees often excluded the lower class, and many people chose to avoid spaces that were clearly demarcated as 'white', 'brown' or 'East Indian'. San Fernando's élite clubs may be divided into two groups: sports clubs that usually provided recreation as well as fielding regular teams; and local branches of international associations that had commercial or charitable objectives. Lower-status clubs also had recreational or sporting facilities, but the majority of blacks and East Indians were not members of them and preferred to relax with friends and kin at home, in bars, or on street corners.

The Naparima, near Paradise Pasture, was San Fernando's most prestigious club. From its foundation after World War I until the early 1950s its membership was predominantly white, both Creole and expatriate, though a handful of coloured professionals eventually gained admission. By the mid-1960s East Indians and light Creoles were prominent among the membership, but dark Negroes were rarely seen around the clubhouse. Respectable blacks tended to join the Promenade Tennis Club off Broadway. Socially mobile East Indians found it easier to get into the Naparima than the Promenade, and their presence at the latter was conspicuous – and encouraged – only on public occasions, such as pre-Lenten dances. At the other extreme was the Palms Club on the Pointe-à-Pierre Road. Unlike the other two, its membership was predominantly black and drawn from one industry – oil; and it provided facilities solely for drinking and dancing.

The clubs discussed so far were essentially Creole dominated, but there were others with East Indian memberships. For example, the Oxford Club, on Lady Hailes Road (Fig. 3.1), was a sports and social club. It was licensed, like the other clubs, but its members were exclusively East Indian and essentially Christian – with a tradition of recruiting from Naparima Boys' School. The Oriental Cricket Club, one of the strongest in southern Trinidad, was also entirely East Indian, but much more representative of the Hindu and Moslem communities.

At the very apex of the social scale, however, multiracialism was rather more evident – especially in the international clubs. Officers of the Southern Chamber of Commerce comprised one East Indian and eight Creoles: the president belonged to one of the town's most prominent Creole families, which owned a large dry-goods store at the bottom of High Street, and his brother was one of San Fernando's two Parliamentarians. Members of the recently formed Lions Club were either doctors, leading businessmen or headteachers: out of 15 members five were East Indian, seven Creole, two Chinese and one was Chinese-East Indian. It was impossible to compile a membership list for San Fernando's rotary club, but I was invited to one of their luncheons. Twenty-four persons were present – 13 white, seven East Indian, three Negro and one 'pass-as-white' – and the majority were well-known business or professional people.

Two conclusions emerge from this study of clubbing. Élite associations remained strongly under the influence of local whites and light browns. In multiracial situations in San Fernando, East Indians were almost invariably Christian. Even more significant still was the endogamy – and solidity – of the principal racial-cultural segments. This lack of inter-marriage in San Fernando is a critical factor, because the endogamous segments provide the social and cultural basis on which manipulators have worked and on which political parties have been built.

Political affiliation

The transformation of Trinidad into the two-party democracy – favoured by British governments – in the years following 1955 has already been described (Chs 2 & 3). Bhadase Maraj's People's Democratic Party (PDP), having lost the 1956 election, evolved into the Democratic Labour Party (DLP) in 1958, and in that year secured a better result, in Trinidad, than the PNM for its federal affiliates in the first and only federal election of the West Indies. Bhadase Maraj was soon replaced as leader by Dr Rudrinath Capildeo, a lecturer in Mathematics at University College, London, in an attempt to overcome the PDP's image as a party of East Indian village headmen and to project the East Indian leadership as comparable to Dr Williams in intellectual capacity. The 1961 election was bitterly fought in an atmosphere of racial hostility, but the PNM won by 20 seats to ten, promptly took Trinidad out of the West Indies Federation in Jamaica's wake, and led the country into independence in 1962 (Lewis 1962). At independence, both political parties were based on the two racial-cultural segments (Bahadoorsingh 1968), the lack of inter-connection between which was the subject of the two preceding sections of this chapter.

Politics transformed these two categories into organised – and competitive – corporate groups. The PNM – under middle-class guidance – created an alliance between the black and brown Creole strata (Oxaal 1968): the DLP, like the former PDP, which was basically the Maha Sabha recast as a political party, was essentially Hindu, but its parliamentary representation involved an unstable coalition of Brahmins with a handful of Moslems and Creoles who broke with the party soon after independence. Political alignments became less culturally and more racially determined after the 1958 federal-election scare for the PNM, to which Premier Williams responded by describing the East Indians as a 'recalcitrant and hostile minority' (Williams 1969, 275). Christian East Indians and Moslems began to withhold their support for the PNM in San Fernando, though not in sufficient numbers to permit DLP victories in 1961 (Malik 1971). Indeed, the PNM recorded clear wins in both the San Fernando seats, where their Creole and Moslem candidates overcame the Moslem and Indo-Chinese contestants put up by the DLP. Nevertheless, Moslem leaders maintained the fiction of being able to deliver their community vote: it gave them bargaining power to secure state funds for their ambitious school-building projects, in addition to the financial and career advantages that came to individuals from access to government patronage.

Voting

Voting behaviour and political opinion in San Fernando around the time of independence can be examined through two sets of data – polling results for the 1961 election and the questionnaire materials collected in 1964. By collapsing the 36 polling districts and 47 census enumeration districts (analysed in Ch. 4) into to 21 common spatial units, statistical comparisons can be made between the percentage voting DLP in each spatial unit in 1961 and the racial and religious categories defined in the 1960 census (Clarke 1972).

The DLP vote was positively but not strongly correlated with Christian East Indians (0.23), Hindus (0.18) and Moslems (0.12). Only the Kendall rank coefficient linking the DLP and the entire East Indian population was as high as 0.28. A larger correlation with the spatial pattern of DLP votes was recorded by Creoles (−0.33), but the sign was negative. This material, too, shows that Creoles and East Indians were politically opposed, and suggests that Creole support for the PNM was stronger than East Indian support for the DLP: indeed for Creoles to vote DLP was tantamount to treason. More surprising was the apparent strength of Christian East Indian support for the DLP, though this was offset by their positive correlation with abstentions (0.25). Moslem association with abstention was even stronger (0.48), thus confirming the increasingly equivocal position of members of this segment *vis-à-vis* the PNM whom they had previously supported, despite the candidacy of a Moslem in San Fernando West. Hindus, however, recorded a correlation with abstention of only 0.07, and for Creoles the coefficient was, of course, negative (−0.20).

Political opinion

The survey data dealing with political opinion have been set out in two tables: the first measures reactions to PNM government, assesses potential manage-

Table 8.2 Political opinions in the San Fernando and Débé samples (1).

Statements or questions	Answers	Creole (%)	*Dougla* (%)	Sample Christian East Indian (%)	Hindu (%)	Moslem (%)	Débé (%)
has the PNM done a good job for people like you?*	good	46.9	23.0	9.8	8.7	7.1	11.0
	fair	48.3	74.4	66.9	61.7	66.7	29.3
	not good	2.4	2.6	18.7	20.8	19.1	37.6
	don't know	2.4	0.0	4.6	8.8	7.1	22.0
	total	100.0	100.0	100.0	100.0	100.0	100.0
which party can best handle the problem of education?†	PNM	83.9	66.6	25.4	21.5	23.8	22.0
	DLP	1.4	10.3	26.6	40.9	35.7	40.3
	don't know	14.7	23.1	48.0	37.6	40.5	37.6
	total	100.0	100.0	100.0	100.0	100.0	100.0
which party can best handle the problem of unemployment?‡	PNM	75.4	48.7	18.8	9.4	10.3	8.2
	DLP	2.4	10.3	36.3	44.9	35.7	44.9
	don't know	22.2	41.0	44.9	45.6	54.0	46.8
	total	100.0	100.0	100.0	100.0	100.0	100.0
voting machines have improved the elections§	agree	85.3	59.0	14.9	14.1	11.1	7.3
	disagree	7.6	33.3	65.7	68.4	77.0	80.7
	don't know	7.1	7.7	19.6	17.4	11.9	11.9
	total	100.0	100.0	100.0	100.0	100.0	100.0
I should vote for a good candidate even if I don't support his party¶	agree	14.2	41.0	39.5	43.6	55.6	28.4
	disagree	83.9	56.4	57.0	53.0	42.1	67.8
	don't know	1.9	2.6	3.5	3.4	2.4	3.8
	total	100.0	100.0	100.0	100.0	100.0	100.0
Dr Capildeo would make a good Prime Minister	agree	14.7	33.3	68.8	79.2	77.8	81.6
	disagree	68.7	48.7	16.8	8.7	5.6	6.5
	don't know	16.6	18.0	14.4	12.1	16.6	11.9
	total	100.0	100.0	100.0	100.0	100.0	100.0
size of samples		211	39	256	149	126	109

**Dougla*–Creole: $\chi^2=9.5$, d.f.$=3$, $P<0.05$; *Dougla*–Christian East Indian: $\chi^2=12.65$, d.f.$=3$, $P<0.01$; Hindu–Débé: $\chi^2=28.21$, d.f.$=3$, $P<0.001$.

†*Dougla*–Creole: $\chi^2=11.82$, d.f.$=2$, $P<0.01$; *Dougla*–Christian East Indian: $\chi^2=27.09$, d.f.$=2$, $P<0.001$; Hindu–Débé: $\chi^2=0.013$, d.f.$=2$, $P>0.99$.

‡*Dougla*–Creole: $\chi^2=13.55$, d.f.$=2$, $P<0.01$; *Dougla*–Christian East Indian: $\chi^2=20.49$, d.f.$=2$, $P<0.001$; Hindu–Christian East Indian: $\chi^2=7.17$, d.f.$=2$, $P<0.05$.

§*Dougla*–Creole: $\chi^2=21.64$, d.f.$=2$, $P<0.001$; *Dougla*–Christian East Indian: $\chi^2=40.26$, d.f.$=2$, $P<0.001$.

¶*Dougla*–Creole: $\chi^2=16.05$, d.f.$=2$, $P<0.001$; *Dougla*–Christian: $\chi^2=0.11$, d.f.$=2$, $P>0.99$, Creole–Débé: $\chi^2=10.87$, d.f.$=2$, $P<0.01$; Christian–Débé: $\chi^2=4.06$, d.f.$=2$, $P>0.10$.

Table 8.3 Political opinions in the San Fernando and Débé samples (2).

Statements or questions	Answers	Creole (%)	*Dougla* (%)	Christian East Indian (%)	Hindu (%)	Moslem (%)	Débé (%)
the two-party system is vital to democracy in Trinidad	agree	95.7	92.3	90.2	83.2	90.5	67.0
	disagree	2.4	7.7	4.3	6.7	3.2	13.8
	don't know	1.9	0.0	5.5	10.1	6.3	19.3
	total	100.0	100.0	100.0	100.0	100.0	100.0
political parties should be based on race	agree	0.5	2.6	0.4	6.0	1.6	10.1
	disagree	98.6	97.4	98.8	92.6	96.8	84.4
	don't know	0.9	0.0	0.8	1.3	1.6	5.5
	total	100.0	100.0	100.0	100.0	100.0	100.0
give your preferred political path for Trinidad in the early 1960s	self-governing	10.1	25.6	44.9	61.7	58.7	52.2
	in Federation	2.8	2.6	6.6	6.0	4.0	5.5
	independent	85.8	71.8	45.7	29.5	36.5	36.7
	don't know	1.4	0.0	2.7	2.7	0.8	4.6
	total	100.0	100.0	100.0	100.0	100.0	100.0
meaning of independence to you	good	33.6	30.8	27.3	21.5	23.0	8.3
	indifferent	49.3	61.5	52.0	71.1	62.7	76.1
	bad	6.6	7.7	17.6	5.4	12.7	11.0
	don't know	10.4	0.0	3.1	2.0	1.6	4.6
	total	100.0	100.0	100.0	100.0	100.0	100.0
which party can better handle the problem of race relations?	PNM	79.6	66.7	16.4	14.8	13.5	4.6
	DLP	2.4	2.6	31.3	36.2	27.8	39.4
	don't know	18.0	30.8	52.3	49.0	58.7	56.0
	total	100.0	100.0	100.0	100.0	100.0	100.0
size of samples		211	39	256	149	126	109

ment of national problems by each party and sets out responses to three key issues – the use of voting machines in the 1961 election, support for candidates versus the parties and the desirability of the Hindu leader of the opposition becoming Prime Minister (Table 8.2); the second looks at the two-party basis for democracy, the racial nature of political parties, the preferred political status for Trinidad and Tobago in the early 1960s, the meaning of independence and the political management of race relations (Table 8.3).

Members of all urban segments agreed that the PNM government had done at least a fair job during the previous eight years: polarisation once again involved the Creoles, 46.9% of whom said the government had done a good job, and Débé villagers, 37.6% of whom replied not good; the three East Indian urban samples gave an intermediate response. Creoles and, to a lesser extent, *Douglas* thought the PNM could best handle the problems of education and unemployment; East Indians tended to disagree, though in the absence of experience of a DLP administration, the majority replied they didn't know. Creoles and *Douglas* believed that voting machines had improved elections, but East Indians – especially those at Débé – disagreed: indeed, most East Indians believed that the PNM had used voting machines to fix the 1961 election. Creoles thoroughly opposed the idea of voting for the candidate rather than the party (83.9%); but leaving aside the Débé sample who also preferred party to candidate, the other East Indian segments were ambivalent. Creoles and *Douglas* thought Dr Capildeo would not make a good Prime Minister: 'Negroes will be made to suffer if ever the DLP with a leader like Dr Capildeo gets into power: there will be racial strife – worse, probably, than in British Guiana' (Creole woman). In contrast, East Indians thought Dr Capildeo would make a good Prime Minister, enthusiasm increasing step by step from Christian East Indians via Hindus and Moslems to Débé residents.

Virtually everyone agreed that the two-party system was vital to democracy, though urban and rural Hindus doubted that Creole hegemony *was* democracy (Table 8.3). There was equally strong opposition – in theory – to political parties being based on race, though a small minority at Débé saw this as perfectly natural. By 1964 hardly anyone wanted Trinidad to have stayed in the West Indies Federation, but there was disagreement over the alternatives. The vast majority of Creoles and *Douglas* opted for independence, Christian East Indians and Creoles were split between favouring independence and continued colonial status, and the majority of Hindus, Moslems and Débé residents preferred continued colonial affiliation to any other outcome; Hindus and Moslems opposed independence under black leadership. Only a minority in all groups viewed independence as good for them: 'Independence is only a name, but does not have a meaning. Negroes are using the word to assert force and power over others' (Moslem man). Hindus, Moslems and Débé residents were notably indifferent to independence, and a small minority of Christian East Indians claimed it was positively bad. Creoles and *Douglas* returned to their more confident vein when assessing the PNM's handling of race relations – which, admittedly, improved very greatly after independence – but the East Indians showed no such enthusiasm and resignedly replied that they didn't know which party could handle the issue better, preferring this response to a favourable comment about the untried DLP.

These results demonstrate a variable but strong polarisation of political opinion between Creoles and East Indians, though East Indian disquiet in 1964 was probably somewhat muted. The *Douglas*, yet again, emerged as an intermediate element, significantly different from the Creoles and Christian East Indians according to the chi-square tests. There was, as the ecological data suggested, a good deal of East Indian solidarity in San Fernando, but the Christians were significantly different from the Moslems and Hindus in certain political attitudes. The material reveals minor shades of distinction between urban Hindus and Moslems, but many clear differences between them and the Débé sample, who, once more, formed the core of the opposition to Creole opinion.

The ecological analysis revealed approximately the same political patterns as the questionnaire, despite the fact that census data for small areas cannot be disaggregated by age and the political correlations are made with the entire segment in question. In general, political attitudes and voting behaviour were similar, though the ecological material was less satisfactory than the questionnaire for differentiating between Hindus, Moslems and Christians, and there was evidence for support for good candidates over ethnic party loyalty among the urban East Indians. 'The racial qualities one sees in the Indians exist only around election times, but in fact they are really nice to get along with . . . There are thousands of Indians who are afraid to let their friends know that they support the PNM. They vote for the party at election times but keep it a secret' (Creole woman).

The ecological results also underestimate the true extent of political opposition between Creoles and East Indians, and this is due to the absence of racial segregation in the town. Moreover, the ecological data emphasise the contribution of Christian East Indian support for the DLP, compared to that of Hindus and Moslems, but this is possibly a statistical artefact, created by the large size of the Christian East Indian population in San Fernando. Racial and cultural segmentation influenced political affiliation, but social differentiation was responsible for only a low level of urban residential segregation, which, in turn, generated small ecological correlations.

This observation immediately suggests another point of view, namely, that class factors may play a considerable rôle in political opinion and affiliation in San Fernando. Cross tabulation of all the previously analysed data on political opinions (Tables 8.2 & 8.3) with occupation, however, shows that the percentages tend to be maintained across class distinctions, with one major contradiction revealed by all samples: the higher the social class position of the respondent the greater the likelihood that responses will deviate from the segmental political norm. Upper-class Creole disaffection from the PNM was particularly noticeable, and many whites and light browns regarded the PNM's anticolonialism as aimed directly at them, especially when Dr Williams (1981, 210–16) declared 'massa day done'. Numerous whites were in the DLP, and a white woman senator, nominated by the party, lived in St Joseph's Village.

Among East Indians, alliance with the PNM strongly characterised the Moslem religious leadership, though several leading Hindus, in varying degrees, were associated with the party. One, an Arya Samaji, was prominent in the Cane Farmers' Association, which he used as a vote bank. Another was prominent in the DLP but was related through marriage to an important East Indian family in the north of Trinidad known for its PNM affiliation. Through this connection he hoped to join the government because it controlled access to resources and patronage. The most interesting of these 'go-betweens' was Nanan (Ch. 6). He was a leading civil servant, a prominent Hindu, despite his Chamar status, and link-man between the PNM and the Hindus in the San Fernando West constituency (which returned the Moslem candidate in 1961). Nanan's political activities merit further consideration, because they illuminate the inter-relationship between a series of key variables – race, religion, class and caste.

A PNM party worker in Port of Spain told me that Nanan had been instructed to 'penetrate the Hindu wall', and this he was well equipped to do. Nanan was politically ambitious, was marginal to the Brahmin-controlled DLP, but in the late 1950s was respected by San Fernando's Hindus because of his high-class position, his organisational ability, and his innovativeness in starting the Ghandi Service League and building the Todd Street *mandir*. By the early 1960s, however, when Nanan gave up the secretaryship to become President of the Service League, it was common knowledge that this organisation was essentially a 'cultural front' for political manipulation, that Nanan was himself

Chairman of Group 16 in San Fernando West constituency, and that, in expectation of receiving honours, he was using his apparent leadership of the Hindus to persuade the government that his political services were indispensable.

In theory, as a high-class member of a low caste, Nanan should have mixed less with Hindus than with members of the Creole segment, to whom he would be acceptable for his class status alone. In practice, Nanan's political standing depended on the support he got, or appeared to get, from Hindus. Nevertheless, Nanan mixed more widely in Negro circles than any other well-known Hindu in San Fernando, and the public occasions with which he was associated as sugar welfare officer, cultural organiser, or political leader were the most skilfully staged multiracial events in town. Of course, these manoeuvres were perfectly obvious to the Hindus, many of whom strongly objected to being used for Nanan's political purposes. A middle-class Brahmin explained that Nanan was 'a Chamar who does Chamar work'. One or two of the Brahmin businessmen who were associated with Nanan were also known to be supporters of the government political party, but they were not called Chamars. It is typical of the sociocultural situation that Creoles should have remained so unaware of Nanan's ineffectualness among Hindus; and equally characteristic of caste that Brahmins should have been excused their faults whereas even the high-class members of low castes, such as Nanan, were criticised for any perceived misdemeanour.

Nanan was listed as a friend by only one person from whom subsample data were elicited. It is significant that this person was a leader of the Moslem community and himself a prominent PNM supporter. Perhaps because I knew both these men through their religious communities – their base of operations – they were fairly helpful to my research. The official line of the local PNM was quite different, however. Although my work was not sabotaged, I was left in no doubt that the project's preoccupation with East Indians met with disapproval. A number of Christian East Indians with PNM connections cut their contact, after indicating enthusiastic interest in the research: and several PNM councillors, both East Indian and Creole, would perform elaborate 'musical chairs' at social events to avoid talking to me.

Conclusion

East Indians and Creoles have been transformed by party processes and organisations from relatively inert colonial corporate categories into organised *de facto* corporate groups. Racially and culturally distinct to start with, these segments were polarised by the struggle for political power between middle-class Creoles (with black electoral support) and Hindus (under Brahmin leadership), and thus remained impervious to the potentially integrating opportunities of inter-marriage, intersegmental friendship and clubbing. However, contact between segments generally did increase as the focus moved away from the most private (marriage) to the most public situation (residence), via friendship and clubbing. The rank order of association among the racial and religious categories was stable from one context to another, however. Christian East

Indians were more 'open' to Creole contacts than were urban Hindus and Moslems, who in turn were less exclusive than Débé villagers. Creoles were essentially assimilationist: they were content to absorb *Douglas* and East Indians who were prepared to associate with them socially, culturally and politically.

The projection of politics on to the plural structure of Trinidad's society has led to a hardening, not softening, of the lines of vertical segmentation. As Lowenthal perceptively commented, 'much of what passes for Indianness . . . is, indeed, a result as well as a cause of East Indian – Creole stress' (Lowenthal 1972, 146). The revival of Indian culture, opening of Hindu schools, and introduction of Hindu teaching under the auspices of the Sanathan Dharma Maha Sabha in the early 1950s were all politically inspired, for Bhadase Maraj aspired to the leadership of the East Indians in a context of quickening West Indian nationalism. Rank-and-file East Indians, too, began to re-emphasise the culture of the indentured Indians so that they would be able to withstand both creolisation and Westernisation and maintain their identity not as peasants in a colonial society but as East Indians in an independent Trinidad.

9 *Conclusion: pluralism in San Fernando and beyond*

San Fernando

The original framework of San Fernando's society during slavery comprised a stratification of Creole cultures whose content and ranking were created through contact among Europeans, Africans and miscegenated coloureds. Differential incorporation of white citizens, free people of colour and black bondsmen projected Creole cultural pluralism into the public domain, where despite slave emancipation in 1834, the introduction of universal adult suffrage in 1946, and political independence in 1962, it persists, *de facto*. However, distinctions between Creole cultural categories lost much of their significance in the context of segmentation produced by the immigration of Indian indentured labour after 1845. East Indians, too, were differentially incorporated as unfree labour (but for five years and not for life, as the slaves had been) and they remained unenfranchised until the introduction of universal adult suffrage.

Indian culture, including religion, language, marriage, the family and caste, though greatly simplified to its basics, persisted intact at independence: indeed, Hinduism and Hindi, Islam and Urdu, all experienced revivals after World War II. East Indians were an endogamous segment or, more accurately, a triad of linked and religiously defined populations external to the Creole stratification. Christian East Indians were closer to Creoles in behaviour and values than to Hindus and Moslems, and half-caste *Douglas* were closer to Creoles than to their East Indian progenitors. Although East Indian culture in San Fernando was continuously renewed by the recruitment of brides from rural areas, and although Hinduism was further strengthened by Brahminism, urban East Indians of all categories were more accommodating to the Creole culture of the 'host' community than were rural East Indians (Hindus), mainly through proximity and ecological circumstance.

Creole–East Indian segmentation was expressed island-wide in segregation and underpinned by urban-rural distinctions. In San Fernando, however, spatial segregation on a racial and religious basis was low, except for whites, largely because the major populations were stratified by occupation (for Moslems and Christian East Indians), by occupation and colour (for Creoles) and by occupation and caste (for Hindus). Although it was impossible to reduce these differing segmental stratifications uniformly to class, Creoles and East Indians enjoyed approximately equal access to housing markets and residential areas.

Census data revealed a continuum in which the Creoles were located at an increasing sociospatial distance from Christian East Indians, Hindus and Moslems, and this was confirmed by the household data, the political opinions of individuals, and other sets of information (Rubin 1962). However, the spatial and survey analyses indicated different degrees of association among the major racial and religious categories. Spatial data, underpinned as they were by the economic conditions typical of a peripheral capitalist society, reflected a marked absence of segregation by segment; conversely the household samples revealed endogamous units that formed social, cultural and microspatial cells in the structure of San Fernando's society (Clarke 1976b).

Segregation of the racial and religious categories in San Fernando increased when the framework of analysis moved from an areal basis to that of the household. There was also a clear and to-be-expected endogamy hierarchy among Hindus: a minority of Hindus married within their caste (49%); the majority married within varna (73%) and to other Hindus (86%); and almost all had East Indian spouses. These regularities were typical of the larger Trinidad scene, as Naipaul confirms: 'marriage between unequal castes has only just ceased to cause trouble; marriage between Hindu and Muslim can still split a family; marriage outside race is unthinkable' (Naipaul 1962, 82).

Although superficially this book confirms that the social structure of San Fernando was reflected in the urban mosaic, the social, cultural and political differences among its populations were more clearly expressed in individual and group behaviour than in their residential distributions. Creoles and East Indians lived in close proximity and shared similar material aspirations, yet their attitudes and behaviour were often poles apart. One of the principal reasons for the difference in the degree of association at the spatial and domestic levels was the dominance of the domestic domain over the public domain, excluding that of politics, in East Indian life. Household and family remained the East Indians' most important reference groups. Parental control of the choice of marriage partner among Hindus, coupled with settlement exogamy, reduced the possibility of interracial marriage implied by residential dispersal, and demonstrated, in an extreme form, the degree to which some East Indians could separate residential and domestic values. This dissociation runs completely against conventional sociogeographic theory which assumes that spatial proximity encourages contact so that marriage distances tend to be short (Beshers 1962). However, although marriage and visiting relationships among Creoles probably took place over short distances, Creoles were not subject to the same range of restrictions as East Indians.

Ecological controls on East Indian behaviour and attitudes in San Fernando were negligible. There was no East Indian ghetto to foster conservative values; no East Indians were so isolated that their values were eroded by Creole opinion; East Indian households were all racially homogeneous, irrespective of location. However, a preoccupation with households and individuals would be misleading if uncorrected by reference to areal data from the census. This study shows that there is clearly value in working simultaneously at different scales, using aggregate and individual data. Scale then ceases to be a trap, and the level of enquiry can be varied to secure new perspectives on the situation, even to incorporate island-wide factors (Clarke 1976b).

East Indians have been creolised by immigration, by indenture, and by the collective experience of about 140 years of settlement in Trinidad. Traditional Indian dress had been modified or lost by 1964, except on ritual occasions, and East Indians in San Fernando usually communicated with one another and with others in Creole English. In the last resort, however, it did not matter whether the Hinduism or Islam they practised were pristine or syncretic, provided that each was *perceived* to embody cultural difference. Despite considerable similarity in household composition among Creoles and East Indians in San Fernando, Hindu and Moslem hostility to mixed marriages ensured racial and cultural endogamy, and guaranteed the biological continuity of the East Indian segment. Politics transformed the East Indian religious categories into functional and organised social groups. The best example of this process was provided by the Sanathan Dharma Maha Sabha, whose Board of Control transformed itself into the management committee of the People's Democratic Party (PDP) in the early 1950s (Malik 1971, 85).

If East Indians were no longer purely 'Indian', by 1964 they had not merged culturally or racially with Creoles. This was particularly true of those East Indians who identified themselves explicitly with their parents' culture, namely Hindus and Moslems; and it was those populations which retained the greatest sense of ethnic identity. By contrast, Christian East Indians were partially acculturated to Creoles behaviourally and attitudinally: and in colour-class terms they were the 'mulattoes' of the East Indian urban population. Although they practised East Indian family systems and dietary habits, they were marginally more likely to have married out of race than out of religion, and their friendships were more oriented towards Creoles than the friendships of Hindus and Moslems. Against this, however, must be set the close kinship links between many Hindus and Christians, especially where conversion had recently occurred.

Urban East Indians have proven more receptive to Creole ways than rural East Indians, and Christian East Indians have provided a bridge however narrow linking East Indians and Creoles – both San Fernando's PNM East Indian mayors in the decade after independence were Presbyterians. *Douglas* have been the outstanding candidates as go-betweens: racially mixed, culturally Creole, but with a residual orientation towards East Indians that facilitated inter-marriage and other social relations, *Douglas* – though small in number – had a potential broker rôle within the urban community, provided that the larger, dominant segments wanted to use it. Throughout the period of the 1950s and 1960s that was unlikely, however, because the most contraposed populations, the urban Creoles and rural Hindus, were locked in an intense struggle to control the government of Trinidad.

Crisis and recovery

The basic concern of this book has been with the decades of decolonisation and the first two years of independence: but how have events in Trinidad worked out since then? The mid-years of the 1960s were characterised by optimistic, neocolonial development based on oil, sugar and light manufacturing, by the 1965 Industrial Stabilisation Act, and by the PNM's comfortable electoral victory in 1966 over the DLP and C. L. R. James's multiracial Workers and Farmers Party. However, material expectations had been raised by party politics, by the media and by independence (Ch. 5) that could not be satisfied either by the modest economic growth or by the government's purchase in the late-1960s of Shell's oil assets and acquisition of a substantial share in the sugar industry. Unemployment, nationally, was at 13%, but was concentrated particularly in the poorer, Creole sections of Port of Spain (Goodenough 1976, 1978). In these conditions Trinidad was ripe for crisis – it came when the international student disturbances of 1968 meshed with US black-power riots and the two, in combination, were projected on to the urban scene by students at the Trinidad campus of the University of the West Indies (Oxaal 1971, Craig 1981b).

The exacerbating rôle played by race and culture in that scene is clearly revealed by a brief examination of segregation in San Fernando in 1970. It was not possible to distinguish Hindus from Moslems using the available census figures, though the enumeration districts were identical to those employed in 1960. Even so, the 1970 census results were pregnant with indications of the tension that then prevailed (Table 9.1). The pattern of the 1960s was for Hindus and Moslems together to maintain their segregation *vis-à-vis* coloured people and blacks, while reducing it *vis-à-vis* Presbyterians; for whites and browns to locate closer together and for blacks to separate themselves not only from Presbyterians, but from white and brown Creoles too. Inclusion of the suburbs around San Fernando reveals even greater increases in white segregation from other categories, but especially from blacks, through the dual processes of town-centre depopulation and surburban growth. Urban discontent at the end of the first decade of sovereignty hinged not on Creole–East Indian stress but on a factor missing from the late 1950s and early 1960s, namely, black disaffection from brown government.

Black-power disturbances, involving student activists at the University of the West Indies at St Augustine and marginalised urban blacks in San Fernando and Port of Spain, led to the declaration of a state of emergency in February 1970, at the very point when unionised oil and sugar workers were entering the fray (Sutton 1983). Part of the national Defence Force sympathetic to black power mutinied, and control was re-established by the government only because of the loyalty of the coastguards and the police (Best 1973). These events split the

Table 9.1 Indices of dissimilarity for racial and religious groups in San Fernando and surrounding suburbs, 1970 (*source:* Census of Trinidad & Tobago, 1970, unpublished data).

	San Fernando town					
Group	White	Mixed	Negro	Oriental religion	Presbyterian East Indian	Chinese
white	—	56.9	68.5	59.0	67.6	68.6
mixed	65.5	—	24.1	26.9	22.5	34.8
Negro	75.8	28.6	—	32.7	34.1	42.7
oriental religion	68.8	31.8	31.5	—	20.4	36.1
Presbyterian East Indian	73.8	27.8	36.1	20.0	—	33.9
Chinese	69.9	38.8	44.4	36.0	35.2	—
	San Fernando and suburbs					

Creole segment along its lines of latent internal cleavage and disrupted the PNM's control of marginalised urban blacks (Oxaal 1971). Fearful that the black-power outburst might engulf them (Nicholls 1971, La Guerre 1974b) and frustrated by a decade of what they saw as Creole gerrymandering of electoral boundaries and fixing of voting machines (Ryan 1972), the DLP took no part in the 1971 election, so that the PNM candidates won all the seats with only a 29% turnout of electors (Greene 1971).

Another major electoral change took place in 1976, with the complete demise of the Hindu DLP and the emergence, as the official opposition, of the United Labour Front (ULF), based on the long-awaited trade union alliance of black oilfield workers and East Indian sugar labourers, an echo of their collaboration in 1937 and 1970 (Ryan 1979). Although the compact was transitory (Ryan, no date), the 1981 election, held after Eric Williams's death, repeated the 1976 contest in its essentials, but with a new party, the Organisation of National Reconstruction, failing to gain a seat, and with George Chambers re-elected Prime Minister by the largest vote the PNM had ever had. Political opposition thereafter devolved upon the Alliance of the ULF, Tapia, a radical-intellectual party, and the Democratic Action Congress, which drew its electoral strength from Trinidad's black dependency, Tobago, and, as the ULF had done six years earlier, from East Indians in the sugar belt.

Since the mid-1970s, the PNM – with some competition from the Organisation for National Reconstruction, has regained control over urban blacks (Ryan *et al.* 1979). It has presented itself to prosperous East Indians of all three segments as the guarantor of the *status quo*, and has ameliorated intersegmental relations by expanding government patronage in the shape of subsidies, tax relief and generation of employment (Sutton 1984, 53). As its electoral position has returned to the high degree of security it enjoyed in the early 1960s, so has its capacity for buying off crises expanded. Discoveries by Amoco of new oil and natural gas deposits under the sea-bed off the Atlantic coast of Trinidad in the early 1970s, backed up by OPEC price increases, have created national wealth undreamed of in 1962 or even 1970: indeed, the GDP increased nine-fold between 1970 and 1980 (Sandoval 1983), and unemployment dropped to 9%. However, oil revenues and reserves are now flagging, and the investments in iron and steel, petrochemicals and fertilisers made by the PNM in the more optimistic 1970s may not sustain national economic growth at the inflated level to which all Trinidadians – especially urbanites – are now accustomed. How will the society cope with increasing urban marginality and with a demographic structure that will probably produce an East Indian majority in the twenty-first century (Harewood 1975)?

Social, cultural and structural pluralism

Lloyd Braithwaite, former Pro-Vice Chancellor of the University of the West Indies at St Augustine and author of the seminal account of Trinidad's society, which focused almost exclusively on Creoles (Braithwaite 1953), has more recently concluded:

> Trinidad, indeed, appears as one of the classic areas of the plural society; perhaps, more correctly, of the plural society arising out of the plantation tradition. Under the umbrella of European imperialism, the labour problems of the plantations were solved by the introduction of African slaves and Indian immigrants. We are left now with a multi-racial, and multi-cultural society that has achieved political independence. Hence, it must seek to resolve our problems, without recourse to outside aid and irrelevant models (Braithwaite 1974, viii).

Smith's three types of pluralism (Ch. 2) – and especially his ideas about social and cultural pluralism – have been explored in this book in the context of an urban community, on a scale much smaller than the state, which is the unit to which they are most easily applied. Until after World War II, San Fernando was a community within a plural society whose differentially incorporated non-white segments were dominated by whites. For more than a century and a half, structural, social and cultural pluralism were locked into a mutually self-sustaining system that could be challenged only by riot and rebellion – or by the uniform incorporation of all Trinidadians on the basis of democracy.

After Trinidad's independence in 1962, however,

> East Indians and Afro-Creoles assert[ed] their equality as individuals and as collectivities, thus manifesting their segmental structure, social pluralism and *de facto* corporate organization. Formerly, under colonial rule, [Trinidad was a complex plurality], since the British differentially incorporated their Afro-Creole and East Indian populations together, while preserving their segmental differences. To some degree, such hierarchic conditions may still persist *sub rosa* in Trinidad, where both the ruling Creole elite and their white allies differ culturally from the black masses on whose support at the polls their power depends (Smith 1984a, 43).

The two major co-ordinate segments – Creole and East Indian – have re-grouped to compete for political power at elections. Their political organisation in the 1950s transformed the two main racial–cultural categories into segmented corporate groups. The electoral competition ostensibly involved East Indian and black, but at the time of independence it crystallised into a struggle between middle-class Creoles and Brahmin Hindus for control of post-imperial society.

To secure their hegemony, however, brown and black politicians appealed to the Christian East Indians and to the Moslem political leadership in the late 1950s and early 1960s, thus recreating in San Fernando the Moslem–Catholic alliance previously orchestrated by Roy Joseph. Similarly, though with little electoral success, the DLP appealed to dissident Moslem and Christian East Indians, and later, after 1970, to defectors from the PNM itself, such as A. N. R. Robinson, the former deputy-leader of Williams's party.

Instability has remained a hallmark of Trinidad society in independence as in colonialism. Unenfranchised blacks and East Indians collaborated in the 1937 disturbances, until the volatile combination of oil and sugar workers was suppressed by the colonial administration which introduced British troops to pacify the colony; black-power disturbances in 1970 exposed the disaffection of the black urban poor, whose marginality had not been ameliorated by Creole, 'brown-man' government, and created circumstances that exploded in the mutiny of the Defence Force (Cross 1972). Since 1975 pursuit of radical politics for the benefit of manual workers has been based on the faltering co-operation of East Indians and blacks in the ULF and, more recently, in the Alliance. Unlike the labour protest of 1937, the multiracial movement has taken place within the democratic framework of elections. The target is no longer a white élite, but the conservative, Creole ruling class.

Much of the aggravation of race relations in Trinidad since 1930 has been due to two causes: socio-economic discontent with the rulers, expressed by black Creoles and East Indians against white rule in 1937 and by black Creoles against brown rule in 1970; and competition between brown and black Creoles and East Indians to succeed to white power in the decolonising years 1946–62. Tensions of the first kind have led to class collaboration between low-status blacks and East Indians; tensions of the second kind have given rise to systematic competition between Creoles and East Indians manipulated by middle-class browns and blacks and by Brahmins at the expense of lower-class blacks and Hindu lower castes. Electoral politics has been the dominant mode of competition since 1946, reinforcing cultural differences at the very time that peasant India was fading as a living memory among East Indians. Indian government scholarships, wearing the sari, and learning Indian dances gave impetus to ethnic identity among opposition Hindus as they competed for national power, and their ethnic identity has flourished yet further since the oil boom with a remarkable elaboration of traditional religious ritual (*jhandi puja*), based upon conspicuous expenditure.

However, if electoral politics have exacerbated the chronic tensions between Creoles and East Indians and revitalised their respective sense of cultural difference, socio-economic tensions associated with unemployment have generated acute social crises of a class type within each of these segments. Hence the significance of 1937 and 1970 as an historical framework for understanding sociocultural and political change in Trinidad as a whole, as well as in San Fernando. Events of 1970 pushed East Indians further away from Creoles but also drew their upper classes closer together when Eric Williams later emerged as the guarantor of the *status quo* and oil wealth created a basis for collaboration (Cross 1978). By 1976 the Hindu political party, the DLP, was quite dead, only the ULF and more recently the Alliance have catered effectively to the East Indian voters: yet the Alliance is committed to promote the collaboration of lower-class blacks and East Indians, a combination that has never succeeded in the past because of their pervasive cultural and social pluralism, and which older and more prosperous East Indians do not want (Ryan 1981, 45–6; La Guerre 1983).

Although neither Creoles nor urban East Indians have overtly adopted racist attitudes to one another, and although there was some similarity of lower-class culture between the segments so far as consensual cohabitation and belief in the supernatural were concerned, it would be unwise to predict convergence or collaboration between Creoles and rural East Indians in the future; Creoles, and East Indians, remain unaware that they have borrowed one another's traits, however minor – for example, Indian food and drumming skills in the case of Creoles. However, objective cultural differences between Creoles and the East

Indian segments may not be crucial to any future political calculus. Urban East Indians are racially proud, biologically exclusive and shed their mixed-race offspring: conversely, black Creoles are racially self-deprecating, biologically inclusive, and incorporate the *Douglas*. The experience of San Fernando's Christian East Indians shows that even if Hinduism and Islam were lost and objective cultural distinctions became nil, racial boundaries would probably remain intact and provide the basis for social segmentation and political alignment.

Transcending the case study

The three modes of pluralism set out by Smith (1974, 1984a & b) may be used to link the San Fernando case study to Trinidad's experience. Differential incorporation in Trinidad created the specific context and circumstances of social and cultural pluralism in San Fernando: slavery destroyed African institutions and moulded the Creole culture of San Fernando, but indenture did not so effectively erode the urban Indian heritage. Both subordinate cultures, neither pristine, were projected by differential incorporation into the public domain in San Fernando and throughout Trinidad, preventing full acculturation of blacks to whites and of East Indians to either. Full acculturation and assimilation are possible only under conditions of uniform incorporation, and may not occur even then if individuals prefer to retain their cultural distinctiveness or if it seems useful to dominant groups – usually political parties – that people should retain their ethnic identity.

Uniform incorporation is indispensable if social pluralism is to be removed from the public domain. However, in modern democracy, cultural pluralism (or alternatively acculturation) yields its full harvest only when governments ensure that racial and cultural factors are not manipulated for political ends. Such circumstances are likely to be rare, especially in those recently independent, successor states that have evolved out of colonial plural societies, whether in the Caribbean or elsewhere, whose experience of liberty and formal equality has been brief.

Direct analogies with the racial–political alignments in Trinidad during the 1960s and 1970s are to be found in its South American neighbours and sibling societies, Surinam and Guyana, where, as Lowenthal notes, parties 'are ethnically constituted, and pressures of self-rule have polarized politics' (1972, 172). Moreover, comparable circumstances occur in Mauritius (Simmons 1982) and Fiji (Mamak 1978). These non-Caribbean cases are different from Trinidad – yet similar to Guyana – in having East Indian demographic majorities, and in *not* having constituency-based, first-past-the-post elections. In none of the five segmental societies above, do East Indians now govern – due largely to the electoral system – though only in Trinidad do they form substantially less than half the population. Solely in Mauritius, where the East Indian majority is largest (68%), have class-based politics, since independence, replaced communal electoral alignments engendered by decolonisation. Everywhere East Indians are seen as interlopers and are not to be allowed to take power as a racial block. In Guyana the East Indian majority are systematically disadvantaged by blacks *de facto*, and elections are rigged using overseas and proxy voting.

So far Trinidad has fared better than many other decolonised plural societies. Having survived the black-power protest of 1970, the Creole PNM has used the country's oil wealth to reduce unemployment and to appease the East Indians while maintaining its own hegemony racially, culturally and politically. However, Trinidad's racial politics remain firmly based on the racial and cultural segmentation that has marginalised the rôle of class affiliation across the lines of segmental cleavage and has left Trinidad dangerously susceptible to race-implicated political crises should oil incomes decline during the next few decades.

Nevertheless, 1970s oil wealth has enabled the favourable economic circumstances recorded in San Fernando in the early 1960s to be extended northwards via the sugar belt to Port of Spain. In the process, San Fernando has become less unique and more characteristic, and the conclusions to be drawn from this study are even more applicable to Trinidadian society today than 20 years ago. These conclusions may be summarised as follows: spatial proximity doesn't necessarily make for integration or reduce social separation in other realms of life, contrary to common belief; household structural similarities do not imply commonality, let alone intimacy, where parental control and racial antipathy ensure endogamy; neither religious conversion nor class mobility erodes racial segmentation; politicisation consequent on independence has made the racial segments more self-conscious and more polarised.

Appendix A *Small-area census data and problems in their analysis*

By manipulating small-area census data enumerated in 1931 and 1960, cartographic and statistical analyses can be used to examine the social and spatial structure of San Fernando at two distinct periods of time: the first, prior to the beginning of the decolonisation period when the racial and cultural hierarchy was still underpinned by self-confident imperialism; the second, just two years short of independence and 14 years after the introduction of universal adult suffrage, by which time non-whites had achieved substantial social mobility.

A number of caveats have to be voiced, however. San Fernando was divided into a mere 20 enumeration districts in 1931, so that the spatial differentiation was rather crude. By 1960, when the town's population had increased by 150%, 47 enumeration districts were available for analysis (leaving aside one unit which contained solely an institutional population) and the potential for subtle spatial and social differentiation was far greater.

The scope of the analysis is limited by the questions asked and data tabulated by the census at the scale of the enumeration district. For example, the 1931 census contains no information on occupation; the 1960 census is devoid of material on housing; and both omit tabulations on small racial minorities such as the Syrians. Moreover, even where the variables appear to be the same in the two censuses, they are often differently defined, though the racial categories, despite changes in nomenclature, remained stable over time.

Three types of analysis are used in the examination of the census data for 1931 and 1960 – mapping of variables, calculation of indices of dissimilarity, and computation of correlation coefficients. Geographers, of course, accept Robinson's (1950) warning that ecological correlations cannot be used as substitutes for individual correlations, but tend to employ the former for their contextual properties, bearing in mind that as areal units get larger, within-group variance increases, between-group variance decreases, and ecological correlations between pairs of variables get larger (Robinson 1950, Woods 1976). A similar but reverse pattern of change affects the calculation of the index of residential dissimilarity between pairs of social groups. As areas get bigger through aggregation, the area-by-area differences between groups tend to decrease and the index of dissimilarity will fall. In practice there is only one spatial unit available in San Fernando for 1931 and 1960 – the enumeration district – and as it remained fairly uniform in population at the two censuses, it does provide a reasonable basis for temporal comparisons of a statistical kind.

Problems associated with the scale and location of the enumeration district graticule affect choropleth maps as well as bivariate statistical analysis. The concept of population density, for example, is clear in the way it applies to any specific district, but as soon as the geographer is free to select the unit for which to calculate the density, he can manipulate, within limits, the result obtained. Another fiction involves the so-called homogeneity of areas. The only way to cope with the complexity of this situation is to ignore variations within small areal units; the smaller the area in question, of course, the more valid is the assumption of uniform conditions.

Appendix B *The sample survey*

An obvious way of escaping the constraints of scale and content implicit in the analysis of census data (Ch. 4) is to carry out sample surveys that are tailor-made to the requirements of the researcher. Individual respondents may be asked a variety of factual and attitudinal questions, and the results can be aggregated to levels appropriate to the particular enquiry. Although sample data do not provide the comprehensive spatial cover achieved by census enumeration, they are a more penetrating and flexible method of investigation, especially where the sample populations, as in this case, have been defined with reference to a social theory – pluralism – tested and confirmed by aggregate analysis.

Preliminary to the questionnaire survey, which focused on segmentation and stratification, religion, family and inter-group association, systematic samples of adults of both sexes were drawn from the electoral roll for San Fernando in 1964. East Indians were distinguished from Creoles on the basis of their names and a random sample of East Indians (531 cases) and a random sample of Creoles (211 cases) were selected. The East Indian sample was divided into three components – Hindu (149 cases), Moslem (126 cases) and Christian (256 cases) – and these can be treated as independent samples. A fifth random sample of adults was taken from the central sections of Débé (109 cases), in the heart of the sugar belt of the Naparimas. Débé was almost entirely East Indian, and four-fifths of its population were Hindu: it was typical of the rural bastions of opposition to the Creole government that were scattered throughout the sugar- and rice-growing areas.

Information about the *Dougla* population of San Fernando (39 cases) – the sixth sample – came to light only as the questionnaire survey among the sample populations proceeded. *Dougla* respondents were collected from surveys among both East Indians and Creoles, and accounted for 6.9% of the combined East Indian and *Dougla* sample populations in San Fernando – compared to 5.1% of the number enumerated in the town in 1931. Since the *Douglas*, as we have seen in Chapter 4, are an important if small element in the racial composition of the town – between Creole and East Indian, but generally cleaving to the Creole – and as it was impossible to secure more satisfactory information about them, this survey material has been used in the analysis. Nevertheless, due regard must be paid to the means by which *Dougla* respondents were obtained and to the small size of the group covered in the survey.

The object in taking these samples was not to estimate the absolute size, say, of the Hindu or Moslem populations, but to measure the frequency distributions of characteristics within and between samples. So, the basic method of analysis involved cross tabulation of the survey data using the Statistical Package for the Social Sciences (SPSS), and, where appropriate, testing the significance of the difference between the samples using chi-square or measuring the sample error per cent.

Appendix C Tables expressing the marital and mating history of household heads and their dependents and showing dependents' relationships to heads

Table C.1 Parental status and marital condition of household heads (NK, not known).

(a) *Creoles*

Mating and parental status	Males					Total		Females					Total	
	<24	25–39	40–54	55–69	70+	No.	%	<24	25–39	40–54	55–69	70+	No.	%
single persons, parental status, NK														
mating now NK	0	0	0	1	0	1	0.6	0	1	0	0	0	1	1.7
mating now	0	4	0	0	0	4	2.5	0	0	0	0	1	1	1.7
not mating now	0	0	1	1	0	2	1.3	0	0	4	1	1	6	10.1
single parents														
mating now NK	0	0	0	0	0	0	0.0	0	1	1	0	0	2	3.4
not mating now	0	0	0	0	0	0	0.0	0	2	0	3	0	5	8.5
mating now	0	1	0	1	0	2	1.3	0	0	0	0	0	0	0.0
consensually wed														
in 1964	0	5	4	1	0	10	6.3	0	0	0	0	0	0	0.0
separated, with children	0	0	2	0	0	2	1.3	0	3	0	1	0	4	6.8
separated, childless	0	1	3	0	0	4	2.5	0	0	0	0	0	0	0.0
separated, not single	0	2	2	1	0	5	3.2	0	0	0	0	0	0	0.0
Married persons														
spouse present	1	30	43	26	0	105	66.5	0	0	0	1	0	1	1.7
spouse absent	0	0	1	0	0	1	0.6	0	1	0	0	0	1	1.7
separated, single	0	1	4	0	0	5	3.2	0	0	7	3	1	11	18.6
separated, not single	0	3	5	2	1	11	6.9	0	0	0	0	0	0	0.0
widowed persons														
widowed, single	0	0	0	3	1	4	2.5	0	1	9	12	3	25	42.4
widowed, not single	0	0	1	1	0	2	1.3	0	0	1	0	1	2	3.4
Total	1	47	71	37	2	158	100.0	0	9	22	21	7	59	100.0

Table C.1 (cont.)

(b) *Douglas*

Mating and parental status	Males					Total		Females					Total	
	<24	25–39	40–54	55–69	70+	No.	%	<24	25–39	40–54	55–69	70+	No.	%
single persons, parental status, NK														
mating now NK	0	0	0	0	0	0	0.0	0	0	0	0	0	0	0.0
mating now	0	0	0	0	0	0	0.0	0	0	0	0	0	0	0.0
not mating now	0	0	0	0	0	0	0.0	0	0	0	0	0	0	0.0
single parents														
mating now NK	0	0	0	0	0	0	0.0	0	0	0	0	0	0	0.0
not mating now	0	0	0	0	0	0	0.0	0	1	0	0	0	1	20.0
mating now	0	0	0	0	0	0	0.0	0	0	0	0	0	0	0.0
consensually wed														
in 1964	0	0	4	0	0	4	17.4	0	0	0	0	0	0	0.0
separated, with children	0	0	0	0	0	0	0.0	0	0	0	0	0	0	0.0
separated, childless	0	0	0	0	0	0	0.0	0	0	0	0	0	0	0.0
separated, not single	0	0	0	0	0	0	0.0	0	0	0	0	0	0	0.0
married persons														
spouse present	1	11	3	1	0	16	69.6	0	1	0	0	0	1	20.0
spouse absent	0	0	0	0	0	0	0.0	0	1	0	0	0	1	20.0
separated, single	0	0	1	0	0	1	4.3	0	1	0	0	0	1	20.0
separated, not single	0	0	1	0	0	1	4.3	0	0	0	0	0	0	0.0
widowed persons														
widowed, single	0	1	0	0	0	1	4.3	0	0	0	1	0	1	20.0
widowed, not single	0	0	0	0	0	0	0.0	0	0	0	0	0	0	0.0
Total	1	12	9	1	0	23	100.0	0	4	0	1	0	5	100.0

Table C.1 (cont.)

(c) *Christian East Indians*

Mating and parental status	Males					Total		Females					Total	
	<24	25–39	40–54	55–69	70+	No.	%	<24	25–39	40–54	55–69	70+	No.	%
single persons, parental status, NK														
mating now NK	1	0	1	0	0	2	0.9	0	0	1	0	0	1	2.3
mating now	0	1	0	0	0	1	0.5	0	1	0	0	0	1	2.3
not mating now	0	1	1	0	1	3	1.5	1	0	2	1	0	4	9.3
single parents														
mating now NK	0	0	0	0	0	0	0.0	0	0	0	0	0	0	0.0
not mating now	0	0	0	0	0	0	0.0	0	1	1	0	0	2	4.7
mating now	0	0	0	0	0	0	0.0	0	0	0	0	0	0	0.0
consensually wed														
in 1964	1	2	1	3	0	7	3.5	0	1	0	0	0	1	2.3
separated, with children	0	0	0	0	0	0	0.0	0	0	0	0	0	0	0.0
separated, childless	0	0	0	0	0	0	0.0	0	0	0	0	0	0	0.0
separated, not single	0	1	1	0	0	2	0.9	0	0	0	0	0	0	0.0
married persons														
spouse present	1	72	62	27	2	164	81.6	0	0	0	0	0	0	0.0
spouse absent	0	0	0	0	0	0	0.0	0	1	1	0	0	2	4.7
separated, single	0	0	3	2	0	5	2.5	0	1	6	2	0	9	20.9
separated, not single	0	4	2	1	1	8	3.9	0	0	0	0	0	0	0.0
widowed persons														
widowed, single	0	1	0	6	1	8	3.9	0	2	12	5	3	22	51.2
widowed, not single	0	1	0	0	0	1	0.5	0	1	0	0	0	1	2.3
Total	3	83	71	39	5	201	100.0	1	8	23	8	3	43	100.0

Table C.1 (cont.)

(d) *Moslems*

Mating and parental status	Males					Total		Females					Total	
	<24	25–39	40–54	55–69	70+	No.	%	<24	25–39	40–54	55–69	70+	No.	%
single persons, parental status, NK														
mating now NK	0	0	0	0	0	0	0.0	0	0	0	0	0	0	0.0
mating now	0	0	0	0	0	0	0.0	0	0	0	0	0	0	0.0
not mating now	0	1	0	0	0	1	0.9	0	0	0	0	0	0	0.0
single parents														
mating now NK	0	0	0	0	0	0	0.0	0	0	0	0	0	0	0.0
not mating now	0	0	0	0	0	0	0.0	0	0	0	1	0	1	5.0
mating now	0	0	0	0	0	0	0.0	0	0	0	0	0	0	0.0
consensually wed														
in 1964	1	2	0	0	0	3	2.7	0	0	0	0	0	0	0.0
separated, with children	0	0	0	0	0	0	0.0	1	0	0	0	0	1	5.0
separated, childless	0	0	0	0	0	0	0.0	0	0	0	0	0	0	0.0
separated, not single	0	0	0	0	0	0	0.0	0	0	0	0	0	0	0.0
married persons														
spouse present	0	28	45	25	1	99	88.4	0	1	2	0	0	3	15.0
spouse absent	0	0	0	0	0	0	0.0	0	0	0	1	0	1	5.0
separated, single	0	2	3	1	0	6	5.4	0	0	1	0	0	1	5.0
separated, not single	0	0	0	0	0	0	0.0	0	0	1	0	0	1	5.0
widowed persons														
widowed, single	0	0	0	1	1	2	1.8	0	1	5	5	1	12	60.0
widowed, not single	0	0	0	0	1	1	0.9	0	0	0	0	0	0	0.0
Total	1	33	48	27	3	112	100.0	1	2	9	7	1	20	100.0

Table C.1 (cont.)

(e) *Hindus*

Mating and parental status	Males					Total		Females					Total	
	<24	25–39	40–54	55–69	70+	No.	%	<24	25–39	40–54	55–69	70+	No.	%
single persons, parental status, NK														
mating now NK	0	0	0	0	0	0	0.0	0	0	0	0	0	0	0.0
mating now	0	1	0	0	0	1	0.7	0	1	0	0	0	0	5.3
not mating now	1	0	1	0	0	2	1.5	0	0	0	0	0	0	0.0
single parents														
mating now NK	0	0	0	0	0	0	0.0	0	0	0	0	0	0	0.0
not mating now	0	0	0	0	0	0	0.0	0	0	0	0	0	0	0.0
mating now	0	0	0	0	0	0	0.0	0	0	0	0	0	0	0.0
consensually wed														
in 1964	0	0	1	2	1	4	2.9	0	0	0	0	0	0	0.0
separated, with children	0	0	0	0	0	0	0.0	0	0	0	0	0	0	0.0
separated, childless	0	0	0	0	0	0	0.0	0	0	0	0	0	0	0.0
separated, not single	0	1	1	1	0	3	2.2	0	0	0	0	0	0	0.0
married persons														
spouse present	1	26	47	20	4	98	72.1	0	0	0	0	0	0	0.0
spouse absent	0	1	0	1	0	2	1.5	0	2	0	0	0	2	10.5
separated, single	0	1	2	0	0	3	2.2	0	0	2	0	0	2	10.5
separated, not single	0	3	9	2	0	14	10.3	0	0	0	0	0	0	0.0
widowed persons														
widowed, single	0	0	1	5	1	7	5.1	0	0	3	11	0	14	73.7
widowed, not single	0	1	1	1	0	2	1.5	0	0	0	0	0	0	0.0
Total	2	33	63	32	6	136	100.0	0	3	5	11	0	19	100.0

Table C.1 (cont.)

(f) *Débé*

Mating and parental status	Males					Total		Females					Total	
	<24	25–39	40–54	55–69	70+	No.	%	<24	25–39	40–54	55–69	70+	No.	%
single persons, parental status, NK														
mating now NK	0	0	0	0	0	0	0.0	0	0	0	0	0	0	0.0
mating now	0	0	0	0	0	0	0.0	0	0	0	0	0	0	0.0
not mating now	0	0	0	0	0	0	0.0	0	0	0	0	0	0	0.0
single parents														
mating now NK	0	0	0	0	0	0	0.0	0	0	0	0	0	0	0.0
not mating now	0	0	0	0	0	0	0.0	0	0	0	0	0	0	0.0
mating now	0	0	0	0	0	0	0.0	0	0	0	0	0	0	0.0
consensually wed														
in 1964	1	0	5	0	0	6	6.0	0	0	0	0	0	0	0.0
separated, with children	0	0	1	0	0	1	1.0	0	0	0	0	0	0	0.0
separated, childless	0	0	0	0	0	0	0.0	0	0	1	0	0	1	11.1
separated, not single	0	0	1	0	0	1	1.0	0	0	0	0	0	0	0.0
married persons														
spouse present	1	31	34	10	2	78	78.0	0	0	0	0	0	0	0.0
spouse absent	0	0	1	0	0	1	1.0	0	0	0	0	0	0	0.0
separated, single	0	1	0	1	1	3	3.0	0	0	0	1	0	1	11.1
separated, not single	0	0	1	1	0	2	2.0	0	0	0	0	0	0	0.0
widowed persons														
widowed, single	0	1	2	2	2	7	7.0	0	2	4	0	0	6	66.6
widowed, not single	0	0	1	0	0	1	1.0	0	0	1	0	0	1	11.1
Total	2	33	46	14	5	100	100.0	0	2	6	1	0	9	100.0

Table C.2 Parental status and marital condition of adults other than the household head (NK, not known).

(a) *Creoles*

Mating and parental status	Males					Total		Females					Total	
	<24	25–39	40–54	55–69	70+	No.	%	<24	25–39	40–54	55–69	70+	No.	%
single persons, parental status, NK														
mating now NK	46	15	1	0	0	62	36.05	37	16	2	0	0	55	17.68
mating now	7	5	0	0	0	12	6.98	3	1	0	0	0	4	1.28
not mating now	70	2	0	1	0	73	42.44	57	5	3	4	0	69	22.19
single parents														
mating now NK	1	1	0	0	0	2	1.16	3	2	0	0	0	5	1.60
not mating now	0	0	0	0	0	0	0.00	0	1	1	1	0	3	0.96
mating now	0	0	0	0	0	0	0.00	0	5	0	0	0	5	1.60
consensually wed														
in 1964	0	1	2	0	0	3	1.74	3	5	5	0	0	13	4.18
separated, with children	0	0	0	0	0	0	0.00	0	0	0	1	1	2	0.64
separated, childless	0	0	1	0	0	1	0.58	0	0	0	0	0	0	0.00
separated, not single	0	0	0	0	0	0	0.00	1	2	1	2	0	6	1.92
married persons														
spouse present	1	5	1	1	2	10	5.81	10	43	44	16	2	115	36.97
spouse absent	0	0	0	1	0	1	0.58	0	3	0	0	0	3	0.96
separated, single	0	5	1	0	0	6	3.49	0	3	2	2	1	8	2.57
separated, not single	0	0	0	0	0	0	0.00	1	1	5	0	0	7	2.25
widowed persons														
widowed, single	0	0	0	0	1	1	0.58	0	0	2	4	8	14	4.50
widowed, not single	0	0	0	1	0	1	0.58	0	0	2	0	0	2	0.64
Total	125	34	6	4	3	172	100.00	112	86	67	30	12	311	100.00

Table C.2 (cont.)

(b) *Douglas*

Mating and parental status	Males <24	Males 25–39	Males 40–54	Males 55–69	Males 70+	Total No.	Total %	Females <24	Females 25–39	Females 40–54	Females 55–69	Females 70+	Total No.	Total %
single persons, parental status, NK														
mating now NK	3	0	0	0	0	3	16.67	0	1	0	0	0	1	3.45
mating now	1	0	0	0	0	1	5.56	0	0	0	0	0	0	0.00
not mating now	7	0	0	0	0	7	38.89	2	0	0	0	0	2	6.90
single parents														
mating now NK	0	0	0	0	0	0	0.00	0	0	0	0	0	0	0.00
not mating now	0	0	0	0	0	0	0.00	0	0	0	0	0	0	0.00
mating now	0	0	0	0	0	0	0.00	0	0	0	0	0	0	0.00
consensually wed														
in 1964	0	1	0	0	0	1	5.56	0	1	2	0	0	3	10.34
separated, with children	0	0	0	0	0	0	0.00	0	0	0	0	0	0	0.00
separated, childless	0	0	0	0	0	0	0.00	0	0	0	0	0	0	0.00
separated, not single	0	0	0	0	0	0	0.00	0	1	1	0	0	2	6.90
married persons														
spouse present	0	4	0	0	0	4	22.22	4	15	1	0	0	20	69.00
spouse absent	0	0	0	0	0	0	0.00	0	0	0	0	0	0	0.00
separated, single	0	0	0	0	0	0	0.00	0	0	1	0	0	1	3.45
separated, not single	0	0	0	0	0	0	0.00	0	0	0	0	0	0	0.00
widowed persons														
widowed, single	1	0	0	0	1	2	11.11	0	0	0	0	0	0	0.00
widowed, not single	0	0	0	0	0	0	0.00	0	0	0	0	0	0	0.00
Total	12	5	0	0	1	18	100.00	6	18	5	0	0	29	100.00

Table C.2 (cont.)

(c) *Christian East Indians*

Mating and parental status	Males					Total		Females					Total	
	<24	25–39	40–54	55–69	70+	No.	%	<24	25–39	40–54	55–69	70+	No.	%
single persons, parental status, NK														
mating now NK	23	2	0	0	0	25	11.47	31	10	1	0	0	42	10.47
mating now	5	1	0	0	0	6	2.75	1	2	1	0	0	4	1.00
not mating now	122	19	1	1	0	143	65.60	101	15	5	1	0	122	30.42
single parents														
mating now NK	0	0	0	0	0	0	0.00	0	0	0	0	0	0	0.00
not mating now	0	1	0	0	0	1	0.46	0	1	0	0	0	1	0.25
mating now	0	0	0	0	0	0	0.00	0	1	0	0	0	1	0.25
consensually wed														
in 1964	0	2	0	0	0	2	0.92	4	2	3	0	0	9	2.24
separated, with children	0	0	0	0	0	0	0.00	0	0	0	0	0	0	0.00
separated, childless	0	0	0	0	0	0	0.00	0	0	1	0	0	1	0.25
separated, not single	0	0	0	0	0	0	0.00	0	2	0	0	0	2	0.50
married persons														
spouse present	8	16	3	0	1	28	12.84	28	93	57	9	0	187	46.63
spouse absent	1	1	1	0	0	3	1.38	1	1	0	0	0	2	0.50
separated, single	0	3	2	0	0	5	2.29	2	3	2	1	0	8	2.00
separated, not single	0	0	1	0	0	1	0.46	0	3	4	0	0	7	1.75
widowed persons														
widowed, single	0	0	1	1	2	4	1.83	0	1	6	4	2	13	3.24
widowed, not single	0	0	0	0	0	0	0.00	0	1	0	1	0	2	0.50
Total	159	45	9	2	3	218	100.00	168	135	80	16	2	401	100.00

Table C.2 (cont.)

(d) *Moslems*

Mating and parental status	Males					Total		Females					Total	
	<24	25–39	40–54	55–69	70+	No.	%	<24	25–39	40–54	55–69	70+	No.	%
single persons, parental status, NK														
mating now NK	14	7	1	0	0	22	13.92	7	1	0	0	0	8	3.15
mating now	1	0	0	0	0	1	0.63	0	0	1	0	0	1	0.39
not mating now	79	13	4	0	0	96	60.76	75	14	0	0	0	89	35.04
single parents														
mating now UK	0	0	0	0	0	0	0.00	1	0	0	0	0	1	0.39
not mating now	0	0	0	0	0	0	0.00	0	1	0	0	0	1	0.39
mating now	0	0	0	0	0	0	0.00	0	0	0	0	0	0	0.00
consensually wed														
in 1964	0	0	0	0	0	0	0.00	1	0	0	0	0	1	0.39
separated, with children	0	0	0	0	0	0	0.00	0	1	0	0	0	1	0.39
separated, childless	0	0	0	0	0	0	0.00	0	0	0	0	0	0	0.00
separated, not single	0	1	1	0	0	2	1.27	1	0	0	0	0	1	0.39
married persons														
spouse present	5	22	4	2	0	33	20.89	21	53	49	8	0	131	51.57
spouse absent	0	0	0	0	1	1	0.63	0	2	0	0	0	2	0.79
separated, single	1	1	0	0	1	3	2.00	0	3	1	2	0	6	2.36
separated, not single	0	0	0	0	0	0	0.00	1	2	2	0	0	5	1.97
widowed persons														
widowed, single	0	0	0	0	0	0	0.00	0	0	0	2	3	5	1.97
widowed, not single	0	0	0	0	0	0	0.00	0	0	2	0	0	2	0.79
Total	100	44	10	2	2	158	100.00	107	77	55	12	3	254	100.00

Table C.2 (cont.)

(e) *Hindus*

Mating and parental status	Males					Total		Females					Total	
	<24	25–39	40–54	55–69	70+	No.	%	<24	25–39	40–54	55–69	70+	No.	%
single persons, parental status, NK														
mating now NK	20	8	1	0	0	29	16.86	8	2	0	0	0	10	3.97
mating now	4	4	0	0	0	8	4.65	2	1	0	0	0	3	1.19
not mating now	82	5	4	0	0	91	52.91	60	7	0	0	0	67	26.59
single parents														
mating now NK	0	0	0	0	0	0	0.00	0	1	0	0	0	1	0.40
not mating now	0	0	0	0	0	0	0.00	0	0	0	0	0	0	0.00
mating now	0	0	0	0	0	0	0.00	1	0	0	0	0	1	0.40
consensually wed														
in 1964	1	1	1	0	0	3	1.74	3	3	1	1	0	8	3.17
separated, with children	0	0	0	0	0	0	0.00	0	1	0	0	0	1	0.40
separated, childless	0	0	0	0	0	0	0.00	0	0	0	0	0	0	0.00
separated, not single	1	0	1	0	0	2	1.16	1	1	2	0	0	4	1.59
married persons														
spouse present	10	17	1	0	1	29	16.86	29	47	39	7	1	123	48.81
spouse absent	0	0	2	1	0	3	1.74	1	0	0	0	0	1	0.40
separated, single	1	2	1	0	0	4	2.33	1	1	0	0	0	2	0.79
separated, not single	0	1	0	0	0	1	0.58	2	4	6	0	0	12	4.76
widowed persons														
widowed, single	0	0	0	2	0	2	1.16	0	4	1	5	3	13	5.16
widowed, not single	0	0	0	0	0	0	0.00	0	0	5	1	0	6	2.38
Total	119	38	11	3	1	172	100.00	108	72	54	14	4	252	100.00

Table C.2 (cont.)

(f) *Débé*

Mating and parental status	Males					Total		Females					Total	
	<24	25–39	40–54	55–69	70+	No.	%	<24	25–39	40–54	55–69	70+	No.	%
single persons, parental status, NK														
mating now NK	14	2	0	0	0	16	15.84	7	0	0	0	0	7	4.02
mating now	0	0	0	0	0	0	0.00	2	0	0	0	0	2	1.15
not mating now	50	4	0	0	0	54	53.47	22	2	0	0	0	24	13.79
single parents														
mating now NK	0	0	0	0	0	0	0.00	2	0	0	0	0	2	1.15
not mating now	0	0	0	0	0	0	0.00	0	0	0	0	0	0	0.00
mating now	0	0	0	0	0	0	0.00	0	0	0	0	0	0	0.00
consensually wed														
in 1964	0	2	0	0	0	2	1.98	2	2	5	0	0	9	5.17
separated, with children	0	0	0	0	0	0	0.00	0	0	0	0	0	0	0.00
separated, childless	0	0	0	0	0	0	0.00	0	0	0	0	0	0	0.00
separated, not single	0	0	0	0	0	0	0.00	0	2	0	0	0	2	1.15
married persons														
spouse present	10	13	1	2	0	26	25.74	28	49	21	5	0	103	59.20
spouse absent	0	1	0	0	0	1	0.99	0	0	0	0	0	0	0.00
separated, single	0	0	1	0	0	1	0.99	3	4	3	0	0	10	5.75
separated, not single	0	0	0	0	0	0	0.00	0	1	1	0	0	2	1.15
widowed persons														
widowed, single	0	0	0	1	0	1	0.99	0	0	3	7	3	13	7.47
widowed, not single	0	0	0	0	0	0	0.00	0	0	0	0	0	0	0.00
Total	74	22	2	3	0	101	100.00	66	60	33	12	3	174	100.00

Table C.3 Population of sample households classified by relationship to household head (HH) and sex of head.

(a) *Creoles*

Categories of kin	Male head No.	Male head %	Female head No.	Female head %	Total No.	Total %
spouses and mates	125	19.53	2	1.02	127	15.19
HH sons	170	26.56	56	28.57	226	27.03
HH daughters	193	30.16	60	30.61	253	30.26
HH sons' sons	2	0.31	0	0.00	2	0.24
HH sons' daughters	4	0.63	2	1.02	6	0.72
HH daughters' sons	9	1.41	7	3.57	16	1.92
HH daughters' daughters	5	0.78	14	7.14	19	2.27
HH daughters' grandchildren	0	0.00	2	1.02	2	0.24
HH mates' sons by others	13	2.03	0	0.00	13	1.56
HH mates' daughters by others	14	2.19	0	0.00	14	1.67
HH mates' daughters' children	7	1.09	0	0.00	7	0.84
HH brothers	3	0.46	2	1.02	5	0.60
HH sisters	2	0.31	8	4.08	10	1.20
HH brothers' children	3	0.47	5	2.55	8	0.96
HH sisters' children	13	2.03	7	3.57	20	2.39
HH brothers' grandchildren	0	0.00	1	0.51	1	0.12
HH mother	5	0.78	5	2.55	10	1.20
HH mothers' mother	0	0.00	1	0.51	1	0.12
HH other matrilateral kin	6	0.94	4	2.04	10	1.20
HH sons' mates and spouses	2	0.31	1	0.51	3	0.36
HH daughters' mates and spouses	1	0.16	2	1.02	3	0.36
adopted persons and issue	12	1.88	5	2.55	17	2.03
unrelated persons and issue	9	1.41	7	3.57	16	1.91
boarders	1	0.16	0	0.00	1	0.12
employees	1	0.16	2	1.02	3	0.36
HH father	1	0.16	0	0.00	1	0.12
HH mates' brothers' children	2	0.31	0	0.00	2	0.24
HH mates' sisters' children	8	1.25	0	0.00	8	0.96
HH mates' sisters and brothers and spouses	3	0.46	1	0.51	4	0.48
HH brothers' spouses and mates	1	0.16	0	0.00	1	0.12
HH mates' mother	8	1.25	0	0.00	8	0.96
HH mates' sisters' children's children	1	0.16	0	0.00	1	0.12
HH sisters' children's children	4	0.63	0	0.00	4	0.48
HH mates' mothers' kin	1	0.16	0	0.00	1	0.12
HH sisters' spouses and mates	1	0.16	1	0.51	2	0.24
HH mates' sons' children	7	1.09	0	0.00	7	0.84
HH mates' fathers' children	1	0.16	0	0.00	1	0.12
HH mates' father	2	0.31	0	0.00	2	0.24
HH mothers' sister	0	0.00	1	0.51	1	0.12
Total	640	100.00	196	100.00	836	100.00

Table C.3 (cont.)

(b) *Douglas*

Categories of kin	Male head No.	Male head %	Female head No.	Female head %	Total No.	Total %
spouses and mates	20	18.87	1	2.22	21	13.91
HH sons	37	34.91	11	24.44	48	31.79
HH daughters	40	37.74	7	15.56	47	31.13
HH sons' sons	0	0.00	3	6.67	3	1.99
HH sons' daughters	0	0.00	4	8.89	4	2.65
HH mates' sons by others	1	0.94	0	0.00	1	0.66
HH mates' daughters by others	2	1.89	0	0.00	2	1.32
HH brothers	0	0.00	4	8.89	4	2.65
HH sisters	0	0.00	8	17.78	8	5.30
HH brothers' children	0	0.00	1	2.22	1	0.66
HH mother	1	0.94	0	0.00	1	0.66
HH sons' mates and spouses	0	0.00	2	4.44	2	1.32
HH sons' mates' children by another	0	0.00	1	2.22	1	0.66
adopted persons and their issue	2	1.89	0	0.00	2	1.32
unrelated persons and their issue	0	0.00	1	2.22	1	0.66
HH father	1	0.94	0	0.00	1	0.66
HH mates' brothers' children	1	0.94	0	0.00	1	0.66
HH mates' sisters and brothers	1	0.94	0	0.00	1	0.66
HH brothers' spouses and mates	0	0.00	2	4.44	2	1.32
Total	106	100.00	45	100.00	151	100.00

Table C.3 (cont.)

(c) *Christian East Indians*

Categories of kin	Male head No.	Male head %	Female head No.	Female head %	Total No.	Total %
spouses and mates	178	16.69	1	0.67	179	14.72
HH sons	322	30.20	44	29.33	366	30.09
HH daughters	372	34.89	42	28.00	414	34.04
HH sons' sons	20	1.87	3	2.00	23	1.89
HH sons' daughters	11	1.03	2	1.33	13	1.07
HH daughters' sons	20	1.87	10	6.67	30	2.46
HH daughters' daughters	17	1.59	17	11.33	34	2.79
HH daughters' grandchildren	1	0.09	0	0.00	1	0.08
HH mates' sons by others	3	0.28	0	0.00	3	0.25
HH mates' daughters by others	6	0.56	0	0.00	6	0.50
HH brothers	12	1.13	6	4.00	18	1.48
HH sisters	6	0.56	9	6.00	15	1.23
HH brothers' children	2	0.19	3	2.00	5	0.41
HH sisters' children	8	0.75	3	2.00	11	0.90
HH mother	7	0.66	1	0.67	8	0.66
HH mothers' mother	1	0.09	0	0.00	1	0.08
HH other matrilateral kin	7	0.66	0	0.00	7	0.58
HH sons' mates and spouses	8	0.75	0	0.00	8	0.66
HH daughters' mates and spouses	11	1.03	2	1.33	13	1.07
HH sons' mates' children by another	1	0.09	1	0.67	2	0.16
other kin of HH's mate and spouse	1	0.09	0	0.00	1	0.08
adopted persons and their issue	4	0.38	3	2.00	7	0.58
unrelated persons and their issue	3	0.28	1	0.67	4	0.33
employees	4	0.38	1	0.67	5	0.41
HH father	2	0.19	0	0.00	2	0.16
other patrilateral kin	2	0.19	0	0.00	2	0.16
HH mates' brothers' children	2	0.19	0	0.00	2	0.16
HH mates' sisters' children and spouses	3	0.28	0	0.00	3	0.25
HH mates' sisters and brothers	9	0.84	0	0.00	9	0.74
HH brothers' spouses and mates	1	0.09	0	0.00	1	0.08
HH mates' mother	5	0.47	0	0.00	5	0.41
HH sisters' children's children	6	0.56	0	0.00	6	0.50
HH sisters' spouses and mates	1	0.09	0	0.00	1	0.08
HH mates' sons' children	2	0.19	0	0.00	2	0.16
HH mates' father	2	0.19	0	0.00	2	0.16
HH mothers' sister	0	0.00	1	0.67	1	0.08
HH sisters' daughters' spouse	1	0.09	0	0.00	1	0.08
HH mates' brothers' spouse	1	0.09	0	0.00	1	0.08
HH mates' sons' spouse	2	0.19	0	0.00	2	0.16
other kin of HH daughters' mate	1	0.09	0	0.00	1	0.08
HH mates' fathers' spouse	1	0.09	0	0.00	1	0.08
Total	1066	100.00	150	100.00	1216	100.00

Table C.3 (cont.)

(d) *Moslems*

Categories of kin	Male head No.	Male head %	Female head No.	Female head %	Total No.	Total %
spouses and mates	107	17.66	4	3.64	111	15.50
HH sons	197	32.51	30	27.27	227	31.70
HH daughters	192	31.68	25	22.73	217	30.31
HH sons' sons	3	0.50	9	8.18	12	1.68
HH sons' daughters	2	0.33	5	4.55	7	0.98
HH daughters' sons	17	2.81	2	1.82	19	2.65
HH daughters' daughters	14	2.31	6	5.45	20	2.79
HH daughters' grandchildren	0	0.00	2	1.82	2	0.28
HH mates' sons by others	10	1.65	0	0.00	10	1.40
HH mates' daughters by others	5	0.83	0	0.00	5	0.70
HH brothers	1	0.17	1	0.91	2	0.28
HH sisters	4	0.66	0	0.00	4	0.56
HH brothers' children	2	0.33	2	1.82	4	0.56
HH sisters' children	1	0.17	2	1.82	3	0.42
HH brothers' grandchildren	0	0.00	1	0.91	1	0.14
HH mother	2	0.33	1	0.91	3	0.42
HH other matrilateral kin	3	0.50	2	1.82	5	0.70
HH sons' mates and spouses	6	0.99	8	7.27	14	1.96
HH daughters' mates and spouses	8	1.32	3	2.73	11	1.54
adopted persons and their issue	5	0.83	0	0.00	5	0.70
unrelated persons and their issue	1	0.17	1	0.91	2	0.28
boarders	1	0.17	0	0.00	1	0.14
employees	1	0.17	1	0.91	2	0.28
HH father	1	0.17	0	0.00	1	0.14
HH mates' brothers' children	2	0.33	0	0.00	2	0.28
HH mates' sisters' children	1	0.17	0	0.00	1	0.14
HH mates' sisters and brothers	7	1.16	0	0.00	7	0.98
HH brothers' spouses and mates	1	0.17	0	0.00	1	0.14
HH mates' mother	3	0.50	0	0.00	3	0.42
HH sisters' sons' spouse	0	0.00	1	0.91	1	0.14
HH sisters' sons' daughter	0	0.00	1	0.91	1	0.14
HH mates' sisters' children's children	2	0.33	0	0.00	2	0.28
HH brothers' children's spouses	0	0.00	1	0.91	1	0.14
HH mates' mothers' brother	1	0.17	0	0.00	1	0.14
HH fathers' mother	1	0.17	0	0.00	1	0.14
HH brothers' spouses' kin	3	0.50	0	0.00	3	0.42
HH mates' brothers' mate	1	0.17	0	0.00	1	0.14
HH daughters' daughters' spouse	0	0.00	1	0.91	1	0.14
HH mothers' sister	0	0.00	1	0.91	1	0.14
HH mates' daughters' mate	1	0.17	0	0.00	1	0.14
Total	606	100.00	110	100.00	716	100.00

Table C.3 (cont.)

(e) *Hindus*

Categories of kin	Male head No.	%	Female head No.	%	Total No.	%
spouses and mates	121	16.80	0	0.00	145	18.44
HH sons	239	33.19	24	36.36	258	32.82
HH daughters	194	26.94	19	28.79	195	24.81
HH sons' sons	25	3.47	1	1.52	28	3.56
HH sons' daughters	15	2.08	3	4.55	21	2.67
HH daughters' sons	12	1.67	6	9.09	12	1.53
HH daughters' daughters	21	2.92	0	0.00	21	2.67
HH daughters' grandchildren	4	0.56	0	0.00	4	0.51
HH mates' sons by others	5	0.69	0	0.00	5	0.64
HH mates' daughters by others	5	0.69	0	0.00	5	0.64
HH brothers	5	0.69	1	1.52	6	0.76
HH sisters	0	0.00	4	6.06	4	0.51
HH brothers' children	6	0.83	0	0.00	6	0.76
HH sisters' children	8	1.11	2	3.03	10	1.27
HH brothers' grandchildren	1	0.14	0	0.00	1	0.13
HH mother	3	0.42	2	3.03	5	0.64
HH sons' mates and spouses	20	2.78	2	3.03	22	2.80
HH daughters' mates and spouses	5	0.69	1	1.52	6	0.76
adopted persons and their issue	5	0.69	0	0.00	5	0.64
unrelated persons and their issue	2	0.28	0	0.00	2	0.25
employees	2	0.28	0	0.00	2	0.25
HH father	4	0.56	0	0.00	4	0.51
HH mates' sisters' children	4	0.56	0	0.00	4	0.51
HH mates' sisters and brothers	3	0.42	1	1.52	4	0.51
HH mates' mother	2	0.28	0	0.00	2	0.25
HH mates' fathers' sister	1	0.14	0	0.00	1	0.13
HH brothers' children's spouses	1	0.14	0	0.00	1	0.13
HH mates' daughters' spouses	1	0.14	0	0.00	1	0.13
HH daughters' sons' spouses	1	0.14	0	0.00	1	0.13
HH daughters' daughters' spouses	1	0.14	0	0.00	1	0.13
HH mates' daughters' children	3	0.42	0	0.00	3	0.38
HH sisters' children's children	1	0.14	0	0.00	1	0.13
Total	720	100.00	66	100.00	786	100.00

Table C.3 (cont.)

(f) *Débé*

Categories of kin	Male head No.	%	Female head No.	%	Total No.	%
spouses and mates	88	14.33	0	0.00	88	13.39
HH sons	217	35.57	10	21.28	227	34.55
HH daughters	170	27.87	18	38.30	188	28.61
HH sons' sons	20	3.29	4	8.51	24	3.65
HH sons' daughters	11	1.80	3	6.38	14	2.13
HH daughters' sons	5	0.82	1	2.13	6	0.91
HH daughters' daughters	6	0.98	4	8.51	10	1.52
HH mates' sons by others	5	0.82	0	0.00	5	0.76
HH mates' daughters by others	3	0.49	0	0.00	3	0.46
HH brothers	1	0.16	0	0.00	1	0.15
HH sisters	5	0.82	0	0.00	5	0.76
HH brothers' children	2	0.33	1	2.13	3	0.46
HH sisters' children	2	0.33	0	0.00	2	0.30
HH brothers' grandchildren	0	0.00	1	2.13	1	0.15
HH mother	8	1.31	0	0.00	8	1.22
HH sons' mates or spouses	16	2.62	0	0.00	16	2.44
HH daughters' mates or spouses	0	0.00	2	4.26	2	0.30
unrelated persons and their issue	2	0.33	2	4.26	4	0.61
HH father	3	0.49	0	0.00	3	0.46
HH mates' brothers' children	3	0.49	0	0.00	3	0.46
HH mates' sisters and brothers	13	2.13	0	0.00	13	1.98
HH mates' sisters' children	1	0.16	0	0.00	1	0.15
HH brothers' spouses and mates	1	0.16	0	0.00	1	0.15
HH mates' mother	5	0.82	0	0.00	5	0.76
HH sons' sons' spouses	4	0.66	0	0.00	4	0.61
HH sons' sons' children	7	1.15	0	0.00	7	1.07
HH brothers' children's spouses	0	0.00	1	2.13	1	0.15
HH mates' sisters' children's children	8	1.31	0	0.00	8	1.22
HH sisters' children's children	2	0.33	0	0.00	2	0.30
HH mates' mothers' kin	1	0.16	0	0.00	1	0.15
HH fathers' brothers' mates	1	0.16	0	0.00	1	0.15
Total	610	100.00	47	100.00	657	100.00

Glossary

aarti Worship.
abir Purple dye thrown on celebrants during *Holi* or *Phagwa*.
agwani Meeting up of bride and groom's fathers at beginning of Arya Samaji wedding.
ashram Hall attached to Hindu temple or *mandir*.
avatar Reincarnation.

barat Motorcade involving Hindu groom and his entourage; traditionally they would have ridden on animals.
bedi Altar.
Bhagwaa Yagya Week-long reading and exposition of Bhagavad-gita.
bhaji Spinach.
black Negro.
Bramha Supreme being of Hinduism, the Creator.
brown Mixed white and black, also called mulatto or coloured.
Buckra Eid Festival that celebrates the story of Abraham and Isaac.

channa Gram.
coloured Mixed white and black, also called mulatto or brown.
comal Hot plate of various sizes on which *roti* is cooked.
Creole A term that distinguishes white, coloured and black West Indians from all others.

dahl Lentil.
dal puja Garlanding ceremony during Hindu wedding.
deota God.
dhantal Percussion instrument.
dharma Religious duty.
dhoti Loincloth giving appearance of loose trousers.
Diwali Festival of lights, devoted to Lakshmi.
Dougla (Hindi, bastard) East Indian and other race mixtures, mostly part-black.

Durga puja Ceremony dedicated to the goddess Durga, usually to ward off sickness.
dwar puja Ceremony of the gateway, Hindu wedding.

Eid ul Fitr Feast that ends the fast of *Ramadan*.

Ganesh God with elephant head (Hindu)
ganghari Traditional long skirt worn by indentured Indian women.
ganja Cannabis or marijuana.
ghee Clarified butter.
Goberdan puja Prayer in honour of Goberdhan Mountain which takes place on Thursday evening of Rameyn Yagya.
gupta dan 'Gift in secret': ritual during Hindu wedding.

hafiz Moslem priest.
haj Pilgrimage (Moslem).
Hanuman Hindu monkey god, also known as Mahabir, whose *puja* is Rōt.
hawan Sacred fire.
holi or *phagwa* Day of secular fun in late March, described in Trinidad as 'Indian carnival'.
Hosay Festival celebrated by Shiites to commemorate the martyrdom of Hossain, Mohammed's grandson, in AD 680.
hypergamy For men, marrying women of a higher status than themselves.
hypogamy For men, marrying women of a lower status than themselves.

iman Moslem priest.

jab molassi Molasses devil.
jahaji bhai Shipmates.
jai mal Abbreviated Hindu wedding ceremony.
jajmani System of economic exchange between castes.

jamaat Congregation of Moslems.
janeo Hindu sacred thread, traditionally taken only by the three (upper) twice-born varnas.
jat Nation or caste.
Jhunam Astamie Birth of Lord Krishna.
jhandi Triangular flag associated with Hindu ritual.
jharay 'Sweeping away' of sickness.
jhula Traditional North Indian blouse, worn by East Indian women.
jouvert Morning of Carnival Monday.
jumbie Creole spirit.

kala pani Black water.
Kali Mai Ceremony devoted to the Hindu goddess Kali.
kalima Faith (Moslem) – there is no god but Allah and Mohammed is his prophet.
kanya dan Virgin giving, part of the Hindu wedding ceremony.
Kartik Nahan Festival of the sea.
katha Reading from sacred Hindu text.
kichree Food served to groom and his kinsmen, eating of which signifies their satisfaction with the wedding gifts they have received.
kitab Reading from the Koran.
kohabar Retiring room used by bride and groom once Hindu wedding is over.
kurta Shirt worn by Hindu men.

La Divina Pastora see *Siparu Mai*.
lagahoo Werewolf (Creole).
Lakshmi Goddess of Light and Knowledge.
lawa Parched rice.
lingam Penis.
lotah Brass vessel for containing water.

mahant Low-caste priest.
maljeu Evil eye (Creole).
mandir Hindu temple.
maro Marriage booth (Hindu).
maro hilai Shaking of the marriage pole by the bridegroom's father as a mark of his satisfaction with the wedding.
matti kore Planting of the marriage pole, Hindu wedding.
maur Crown worn by the groom at Hindu wedding.
milap Meeting up of bride and groom's fathers and their entourages prior to the Hindu wedding.
mulatto Coloured, white–black, also called brown.
mullah Moslem spiritual leader.

namaaz Prayers (Moslem).
Nau Barber caste. Nau's wife (often a rôle played by a woman of another caste) is important in the marriage ritual at Hindu weddings.
Negro Black.

obeah man Black magic practitioner (Creole).
Ogun Shango god of War, syncretised with St Michael.
ojha man Hindu practitioner in black magic.
ole mas 'Old mask': fancy dress competition during *jouvert*.
oronhi Veil worn by married Moslem and Hindu women.

panchayat Caste council.
parata roti Special *roti* of Moslems.
patna Spirit, believed in by Hindus, that makes girls barren.
patra Horoscope used by Hindu priests.
pau puja Gifting ceremony during Hindu wedding.
persad Sacred food.
pirah Wedding bench (Hindu) made from a single piece of wood and unjointed.
puja Prayer ceremony devoted to Hindu deity.
pundit Hindu priest.

Radha Consort of Krishna.
rakaat Genuflection during Moslem prayers.
Ram Naumi Celebration of Rama's birth.
Ramadan Month-long fast during daylight hours enjoined upon orthodox Moslems.
Ramanandi Panthi Sect focusing on the divinity of Lord Rama.
Rameyn satsang Sung version of *Ramayana*.

Rameyn Yagya Week-long reading and exposition of the Ramayana.
Ramlila Rama pageant.
Rōt Hanuman puja.
roti Unleavened bread.
roza Fasting (Moslem).

saddhu Holy man.
Sanathan Dharma Maha Sabha Hindu socioreligious organisation: the most orthodox of Hindu sects.
Sandhya puja Daily prayers of the 'twice-born' Hindus.
sanskritise Living by the tenets of orthodox Hinduism as set down in sanskrit.
saptapadi Seven steps around the sacred fire that seal the Hindu marriage.
Satnaryn God of Truth.
Shango Afro-Christian cult named after Yoruba god of war.
siballa best man, Hindu wedding.
sindoor Red powder rubbed into the central hair parting of Hindu woman to indicate her married status.
Siparu Mai Siparia Mother or *La Divina Pastora*, the black Virgin of Siparia, equated by Hindus with *Kali Mai*.
sirdar Overseer.
Siva God the destroyer.
Siw Ratri Celebration of Siva's birth.
soucouyant Vampire (Creole).
Suruj Puran Sun god.

tabla Drum played with the fingers.
tadjahs Wood, paper and tinsel representations of the tomb of the martyr Hossain, constructed for Hosay, to commemorate the martyrdom.
tarriah Brass tray.
tilak Betrothal (Hindu).

Vishnu God of the Preserver.

zakat Compulsory charity (Moslem).

Bibliography

Ahsan, S. R. 1963. *East Indian agricultural settlements in Trinidad: a study in cultural geography*. University of Florida, unpublished PhD dissertation.

Allum, D. 1973. Legality vs. morality: a plea for the Lt. Raffique Shah. In *The aftermath of sovereignty: West Indian perspectives*, D. Lowenthal and L. Comitas (eds), 331–48. Garden City, New York: Doubleday.

Angrosino, M. 1974. *Outside is death: community organization, ideology and alcoholism among East Indians of Trinidad*. Winston-Salem, North Carolina: Overseas Research Centre, Wake Forest University.

Angrosino, M. V. 1975. V. S. Naipaul and the colonial image. *Carib. Q.* **21**(3), 1–11.

Angrosino, M. V. 1976. Sexual politics in the East Indian family in Trinidad. *Carib. Stud.* **16**(1), 44–66.

Anthony, M. 1963. *The games were coming*. London: André Deutsch.

Anthony, M. 1965. *The year in San Fernando*. London: André Deutsch.

Anthony, M. and A. Carr 1975. *Trinidad and Tobago*. London: André Deutsch.

Augelli, J. and H. W. Taylor 1960. Race and population patterns in Trinidad. *Ann. Assoc. Am. Geogs.* **20**(2) 123–38.

Bahadoorsingh, K. 1968. *Trinidad electoral politics: the persistence of the race factor*. London: Institute of Race Relations.

Baksh, I. J. 1979. Stereotypes of Negroes and East Indians in Trinidad: a re-examination. *Carib. Q.* **25**(1 & 2), 52–71.

Basdeo, S. 1981. The role of the British labour movement in the development of labour organisation in Trinidad 1919–29. *Soc. Econ. Stud.* **30**, 21–41.

Basdeo, S. 1982a. The 1934 labour disturbances in Trinidad: a case study in colonial labour relations. In *East Indians in the Caribbean*, B. Brereton and W. Dookeran (eds), 49–70. New York, London and Nendeln: Kraus International.

Basdeo, S. 1982b. The role of the British labour movement in the development of labour organization in Trinidad 1929–38. *Soc. Econ. Stud.* **31**, 40–73.

Bell, R. R. 1970. Marriage and family differences among lower class Negro and East Indian women in Trinidad. *Race* **12**, 59–73.

Benedict, B. 1961. *Indians in a plural society: a report on Mauritius*. London: Her Majesty's Stationery Office.

Benedict, B. 1962. Stratification in plural societies. *Am. Anthrop.* **64**, 1235–46.

Benedict, B. 1965. *Mauritius: problems of a plural society*. London: Pall Mall Press.

van den Berghe, P. 1967. *Race and racism: a comparative perspective*. London: Wiley.

van den Berghe, P. 1973. Pluralism. In *Handbook of social and cultural anthropology*, J. Honigman (ed.), 959–78. Chicago: Rand-McNally.

Beshers, J. M. 1962. *Urban social structure*. New York: Free Press of Glencoe.

Best, L. 1965. Chaguaramas to slavery? *New World, Dead Season*, 43–70.

Best, L. 1973. The February revolution. In *The aftermath of sovereignty*, D. Lowenthal and L. Comitas (eds), 306–29. Garden City, New York: Doubleday.

Bisnauth, D. A. 1982. Religious pluralism and development in the Caribbean. *Carib. J. Relig. Stud.* **4**(2), 17–33.

Blouet, B. W. 1976. Land policies in Trinidad 1838–50. *J. Carib. Hist.* **9**, 43–59.

Bowen, N. P. and B. G. Montserin 1948. *Colony of Trinidad and Tobago census album*. Port of Spain: Government Press.

Braithwaite, F. S. 1980. Race, social class and the origins of occupational elites in Trinidad and Tobago. *Bol. Estud. Latinoam. Caribe* **28**, 13–30.

Braithwaite, L. 1953. Social stratification in Trinidad. *Soc. Econ. Stud.* **2**, 5–175.

Braithwaite, L. 1954. The problem of cultural integration in Trinidad. *Soc. Econ. Stud.* **3**, 82–96.

Braithwaite, L. 1960. Social stratification and social pluralism. In *Social and cultural pluralism in the Caribbean*, V. Rubin (ed.). *Ann. N.Y. Acad. Sci.* **83**, 816–36. New York: New York Academy of Sciences.

Braithwaite, L. 1974. Foreword. In *Calcutta to Caroni*, J. La Guerre (ed.), vii–viii. London: Longman.

Braithwaite, L. and G. W. Roberts. 1967. Mating patterns and prospects in Trinidad. *Centr. Stat. Off. Res. Paps.* **4**, 119–27.

Brereton, B. 1974a. The experience of indentureship, 1845–1917. In *Calcutta to Caroni*, J. La Guerre (ed.), 25–38. London: Longman.

Brereton, B. 1974b. The foundations of prejudice. Indians and Africans in 19th century Trinidad. *Carib. Iss.* **1**, 15–28.

Brereton, B. 1979. *Race relations in colonial Trinidad*. Cambridge, London and New York: Cambridge University Press.

Brereton, B. and W. Dookeran (eds) 1982. *East Indians in the Caribbean*. New York and London: Kraus International.

Buraway, M. 1974. Race, class and colonialism. *Soc. Econ. Stud.* **23**, 521–50.

Camejo, A. 1971. Racial discrimination in employment in the private sector in Trinidad and Tobago: a study of the business elite and the social structure. *Soc. Econ. Stud.* **20**, 294–318.

Caribbean Dialogue 1976. At the crossroads: a special double issue on the political situation in Trinidad and Tobago. **2**(3 & 4).

Caribbean Issues 1976. Perspectives on East Indians in the Caribbean. **2**(3).

Caribbean Quarterly 1956. Trinidad carnival issue. **4**, 173–318.

Carmichael, G. 1961. *The history of the West Indian islands of Trinidad and Tobago*. London: Alvin Redman.

Carr, A. 1953. A rada community in Trinidad. *Carib. Q.* **3**(1), 35–54.

Carr, A. T. No date. Trinidad calypso is unique folk culture. Unpublished.

Carrington, E. 1967. The post-war political economy of Trinidad and Tobago. *New World Q.* **4**(1), 45–67.

Carrington, E. 1968. Industrialization in Trinidad and Tobago since 1950. *New World Q.* **4**(2), 37–43.

Clarke, C. G. 1967. Caste among Hindus in a town in Trinidad: San Fernando. In *Caste in overseas Indian communities*, B. M. Schwartz (ed.), 165–99. San Francisco: Chandler.

Clarke, C. G. 1971. Residential segregation and intermarriage in San Fernando, Trinidad. *Geogr. Rev.* **61**, 198–218.

Clarke, C. 1972. The political ecology of a town in Trinidad. In *International Geography 1972*, W. Peter Adams and F. M. Helleiner (eds), 798–801. Montreal.

Clarke, C. G. 1973. Pluralism and stratification in San Fernando, Trinidad. In *Social patterns in cities*, B. D. Clark and M. B. Gleave (compilers), 53–70. Special Publication, no. 5. London: Institute of British Geographers.

Clarke, C. 1975. *Kingston Jamaica: urban growth and social change, 1692–1962*. Berkeley, Los Angeles and London: University of California Press.

Clarke, C. 1976a. Scale factors and ethnic patterns in Trinidad. *Geogr. Soc. J.* (University of Liverpool) **4**, 30–4.

Clarke, C. 1976b. Aggregate data and the behaviour of individuals: perspectives from a multi-racial urban community. In *Population at microscale*, L. A. Kosinski and J. W. Webb (eds), 149–72. New Zealand Geographical Society.

Clarke, C. 1984. Pluralism and plural societies: Caribbean perspectives. In *Geography and ethnic pluralism*, C. Clarke, D. Ley and C. Peach (eds), 51–86. London: Allen & Unwin.

Collens, J. H. 1888. *Guide to Trinidad*. London: Elliot Stock.
Comma, C. N. (ed.) 1966. *Who's who in Trinidad and Tobago*. Port of Spain.
Craig, S. 1981a. Sociological theorizing in the English-speaking Caribbean: a review. In *Contemporary Caribbean: a sociological reader*, S. Craig (ed.), vol. 2, 143–80. Port of Spain: Susan Craig.
Craig, S. 1981b. Background to the 1970 confrontation in Trinidad and Tobago. In *Contemporary Caribbean: a sociological reader*, S. Craig (ed.), vol. 2, 385–423. Port of Spain: Susan Craig.
Crooke, W. 1896. *Tribes and castes of the North-Western Provinces and Oudh* (4 vols). Calcutta: Office of the Superintendent of Government Printing.
Crooke, W. 1897. *The North West Provinces of India*. London: Methuen.
Cross, M. 1972. *The East Indians of Guyana and Trinidad*. Report no. 13. London: Minority Rights Group; new edition 1980.
Cross, M. 1978. Colonialism and ethnicity: a theory and comparative caste study. *Ethn. Rac. Stud.* **1**, 37–59.
Cross, M. and A. Schwartzbaum, 1969. Social mobility and secondary school selection in Trinidad and Tobago. *Soc. Econ. Stud.* **18**, 189–207.
Crowley, D. J. 1957. Plural and differential acculturation in Trinidad. *Am. Anthrop.* **59**, 817–24.
Crowley, D. J. 1958. Calypso, Trinidad carnival songs and dances. *Dance Notat. Rec.* **9**, 3–7.
Crowley, D. J. 1961. Cultural association in a multi-racial society. In *Social and cultural pluralism in the Caribbean*, V. Rubin (ed.). *Ann. N.Y. Acad. Sci.* **83**, 850–4. New York: New York Academy of Sciences.
Cumpston, I. M. 1953. *Indians overseas in British territories, 1834–53*. Oxford: Oxford University Press.
Current Sociology 1959. Caste: a trend report and bibliography **8**(3), 135–83.

Davids, L. 1964. The East Indian family overseas. *Soc. Econ. Stud.* **13**, 383–96.
Delf, G. 1963. *Asians in East Africa*. London, New York and Nairobi: Oxford University Press.
Deosaran, R. 1981. Some issues in multiculturalism: the case of Trinidad and Tobago in the post-colonial era. *Ethn. Stud.* **3**, 199–225.
Despres, L. A. 1967. *Cultural pluralism and national politics in British Guiana*. Chicago: Rand McNally.
Dew, E. 1978. *The difficult flowering of Surinam: ethnicity and politics in a plural society*. The Hague: Martinus Nijhoff.
Dookeran, W. 1974. East Indians and the economy of Trinidad and Tobago. In *Calcutta to Caroni*, J. La Guerre (ed.), 69–83. London: Longman.
Durbin, M. A. 1973. Formal changes in Trinidad Hindi as a result of language adaptation. *Am. Anthrop.* **75**, 1290–304.

Farrell, T. M. A. 1978. The unemployment crisis in Trinidad and Tobago: its current dimensions and projections to 1985. *Soc. Econ. Stud.* **27**, 117–52.
Forbes, R. 1985 *Hindu organizational function and dysfunction in an alien society*. Trinidad: Vedante Society.
Franck, H. A. 1923. *Roaming through the West Indies*. New York: Century Co.
Freilich, M. 1961. Serial polygyny, Negro peasants and model analysis. *Am. Anthrop.* **63**, 955–73.
Froude, J. A. 1888. *The English in the West Indies*. London: Longman, Green.
Furnivall, J. S. 1948. *Colonial policy and practice: a comparative study of Burma and Netherlands India*. London: Cambridge University Press.

Gastmann, A. L. 1968. *The politics of Surinam and the Netherlands Antilles*. Rio Piedras, Puerto Rico: Institute of Caribbean Studies.

Genovese, E. D. 1971. *In red and black*. London: Penguin.

Gerber, S. N. (ed.) 1968. *The family in the Caribbean*. Rio Piedras, Puerto Rico: Institute of Caribbean Studies.

Gerth, H. H. and C. W. Mills (eds) 1970). *From Max Weber: essays in sociology*. London: Routledge & Kegan Paul.

Giacottino, J. C. 1969. L'Économie Trinidadienne. *Cahiers Outre-Mer* **22**, 113–60.

Giacottino, J. C. 1977. *Trinidad et Tobago: étude geographique* (3 vols). L'Université de Bordeau III, unpublished doctoral thesis.

Gillion, K. L. 1962. *Fiji's Indian migrants*. Melbourne, London, Wellington and New York: Oxford University Press.

Glasgow, R. A. 1970. *Guyana: race and politics among Africans and East Indians*. The Hague: Martinus Nijhoff.

Glazier, S. D. 1980. Pentacostal exorcism and modernization in Trinidad. In *Perspectives on pentacostalism*, S. D. Glazier (ed.), 67–80. Washington D.C.: University Press of America.

Glazier, S. D. 1982. African cults and Christian churches in Trinidad: the spiritualist Baptist case. *J. Relig. Th.* **29**(2), 17–25.

Glazier, S. D. 1983a. Caribbean pilgrimages: a typology. *J. Sci. Stud. Relig.* **22**, 316–25.

Glazier, S. D. 1983b. Cultural pluralism and respectability in Trinidad. *Ethn. Rac. Stud.* **6**, 351–55.

Gomes, A. 1974. *Through a maze of colour*. Port of Spain: Key Caribbean Publications.

Goodenough, S. S. 1976. *Race status and residence: Port of Spain, Trinidad*. University of Liverpool, unpublished PhD thesis.

Goodenough, S. S. 1978. Race, status and urban ecology in Port of Spain, Trinidad. In *Caribbean social relations*, C. G. Clarke (ed.), 17–45. Monograph Series no. 8. Liverpool: Centre for Latin-American Studies, University of Liverpool.

Grant, K. G. 1923. *My missionary memories*. Halifax, Nova Scotia: Imperial Publishing.

Greene, J. E. 1971. An analysis of the general elections in Trinidad and Tobago 1971. In *Reading in government and politics of the West Indies*, T. Munroe and R. Lewis (eds), 136–45. Jamaica: Department of Government, University of the West Indies.

Green, H. B. 1964. Socialization values in the Negro and East Indian subcultures of Trinidad. *J. Soc. Psychol.* **64**, 1–20.

Green, H. B. 1965. Values of negro and East Indian children in Trinidad. *Soc. Econ. Stud.* **14**, 204–16.

Greene, J. E. 1974. *Race vs politics in Guyana*. Jamaica: Institute of Social and Economic Research, University of the West Indies.

Haraksingh, K. 1979. Control and resistance among overseas Indian workers: a study of labour on the sugar plantations of Trinidad. Paper presented at the 43rd Congress of Americanists, British Columbia 1979; later published in *J. Carib. Hist.* **14**, 1–17.

Harewood, J. No date. *Employment in Trinidad and Tobago 1960*. Jamaica: Institute of Social and Economic Research, University of the West Indies.

Harewood, J. 1967. Population growth in Trinidad and Tobago in the twentieth century. *Centr. Stat. Off. Res. Paps* **4**, 69–94.

Harewood, J. 1971. Racial discrimination in employment in Trinidad and Tobago *Soc. Econ. Stud.* **20**, 267–93.

Harewood, J. 1975. *The population of Trinidad and Tobago*. Paris: CICRED.

Harris, J. H. 1910. *Coolie labour in the British Crown Colonies and protectorates*. London.

Harrison, D. 1979. The changing fortunes of a Trinidad peasantry. In *Peasants, plantations and rural communities in the Caribbean*, M. Cross and A. Marks (eds). Guildford and

Leiden: Department of Sociology, University of Surrey and Royal Institute of Linguistics and Anthropology, Leiden.
Harvey, D. W. 1973. *Social justice and the city*. London: Edward Arnold.
Henriques, F. 1953. *Family and colour in Jamaica*. London: Eyre & Spottiswoode.
Henry, F. 1965. Social stratification in an Afro-American cult. *Anthropol. Q.* **38**(2), 72–8.
Henshall, J. D. 1984. Gender versus ethnic pluralism in Caribbean agriculture. In *Geography and ethnic pluralism*, C. Clarke, D. Ley and C. Peach (eds), 173–92. London: Allen & Unwin.
Herskovits, M. J. and F. S. Herskovits 1947. *Trinidad village*, 1964 edn. New York: Octagon Books. (First published, New York: Alfred Knopf.)
Higman, B. W. 1984. *Slave populations of the British Caribbean, 1807–34*. Baltimore: Johns Hopkins University Press.
Hill, E. 1972. *The Trinidad carnival*. Austin and London: University of Texas Press.
Hutton, J. H. 1961. *Caste in India*, 3rd edn. London: Oxford University Press.
Hyman, R. 1971. *Lower class families: the culture of poverty in Negro Trinidad*. New York, London and Toronto: Oxford University Press.

Jacobs, W. R. 1977. The politics of protest in Trinidad: the strikes and disturbances of 1937. *Carib. Stud.* **17**(1–2), 5–54.
James, C. L. R. No date. *Party politics in the West Indies*. San Juan, Trinidad: Vedic Enterprises.
Jayawardena, C. 1960. Marital stability in two Guianese sugar estate communities *Soc. Econ. Stud.* **9**, 76–100.
Jayawardena, C. 1962. Family organization in plantations in British Guiana. *J. Compar. Sociol.* **3**, 43–65.
Jayawardena, C. 1966. Religious belief and social change. *Compar. Stud. Soc. Hist.* **8**, 211–40.
Jayawardena, C. 1968. Migration and social change: a survey of Indian communities overseas. *Geogr. Rev.* **58**, 426–49.
Jha, J. C. 1973. Indian heritage in Trinidad, West Indies. *Carib. Q.* **19**(2), 28–50.
Jha, J. C. 1974. The Indian heritage in Trinidad. In *Calcutta to Caroni*, J. La Guerre (ed.), 1–24. London: Longman.
Jha, J. C. 1976a. The Hindu sacraments (*rites de passage*) in Trinidad and Tobago. *Carib. Q.* **22**(1), 40–52.
Jha, J. C. 1976b. The Hindu festival of Divali in the Caribbean. *Carib. Q.* **22**(1), 53–61.
Jha, J. C. 1982. The background of the legalisation of non-Christian marriages in Trinidad and Tobago. In *East Indians in the Caribbean*, B. Brereton and W. Dookeran (eds), 117–39. New York and London: Kraus International.
Johnson, H. 1971. Immigration and the sugar industry in Trinidad during the last quarter of the 19th century. *J. Carib. Hist.* **3**, 28–72.
Johnson, H. 1972. The origins and early development of cane farming in Trinidad, 1882–1906. *J. Carib. Hist.* **5**, 46–74.
Johnson, H. 1973. Barbadian immigrants in Trinidad, 1870–97. *Carib. Stud.* **13**(3), 5–30.
Joseph, E. L. 1838. *The history of Trinidad*, reprinted in Cass Library of West Indian Studies, no. 13, 1970. London: Cass.

Khan, I. 1961. *The jumbie bird*. London: Macgibbon & Kee.
King, A. D. 1976. *Colonial urban development: culture, social power and environment*. London: Routledge and Kegan Paul.
Kingsley, C. 1900. *At last: a Christmas in the West Indies*, 3rd edn. London: Macmillan.
Kirpalani, M. J., M. G. Sinanan, and L. F. Seukeran (eds) 1945. *Indian centenary review: one hundred years of progress 1845–1945. Trinidad, B.W.I.* Port of Spain: Indian Centenary Review Committee.

Klass, M. 1960. East and West Indian: cultural complexity in Trinidad. In *Social and cultural pluralism in the Caribbean*, V. Rubin (ed.). *Ann. N.Y. Acad. Sci.* **83**, 855–61. New York: New York Academy of Sciences.

Klass, M. 1961. *East Indians in Trinidad*. New York: Columbia University Press.

Knopp, S. F. 1961. Shanty town: the study of a newly formed outcast group in Trinidad. Faculty of Political Science, Columbia University, unpublished MA thesis.

Kuper, H. 1960. *Indian people in Natal*. Natal: Natal University Press.

Kuper, L. 1969. Ethnic and racial pluralism: some aspects of polarization and depluralization. In *Pluralism in Africa*, L. Kuper and M. G. Smith (eds), 459–87. Berkeley, Los Angeles and London: University of California Press.

La Guerre, J. 1972. The general elections of 1946 in Trinidad and Tobago. *Soc. Econ. Stud.* **21**, 184–203.

La Guerre, J. 1974a. The East Indian middle class today. In *Calcutta to Caroni*, J. La Guerre (ed.), 98–107. London: Longman.

La Guerre, J. 1974b. Afro-East Indian relations in Trinidad and Tobago. *Carib. Iss.* **1**, 49–61.

La Guerre, J. 1982. *The politics of communalism: the agony of the left in Trinidad and Tobago 1930–1955*. Trinidad: Pan Caribbean.

La Guerre, J. 1983. The general elections of 1981 in Trinidad and Tobago. *J. Commonwth Compar. Polit.* **21**, 133–51.

La Guerre, J., B. Samaroo and G. Sammy no date. *East Indians and the present crisis*.

Lamur, H. E. 1973. *The demographic evolution of Surinam 1920–1970*. The Hague: Maritinus Nijhoff.

Laurence, K. O. 1971a. *Immigration into the West Indies in the nineteenth century*. Barbados: Caribbean Universities Press.

Laurence, K. O. 1971b. East Indian indenture in Trinidad. *Carib. Q.* **17**(1), 34–47.

Lemon, A. 1980. The Indian communities of East Africa and the Caribbean. In *Studies in overseas settlement and population*, A. Lemon and N. C. Pollock (eds), 225–41. London and New York: Longman.

Lemon, A. 1984. The Indian and coloured elections: co-optation rejected? *S. Afr. Int.* **15**(2), 84–107.

Lewis, G. K. 1962. The Trinidad and Tobago general elections of 1961. *Carib. Stud.* **2**(2), 2–30.

Lewis, M. W. 1978. Yoruba religion in Trinidad: transfer and reinterpretation. *Carib. Q.* **24**(3 & 4), 18–32.

Lieber, M. 1981. *Street scenes: Afro-American culture in urban Trinidad*. Cambridge, Mass: Schenkman.

Lobb, J. 1940. Caste and class in Haiti. *Am. J. Sociol.* **46**, 23–34.

Lowenthal, D. 1972. *West Indian societies*. London: Oxford University Press.

MacDonald, J. S. and L. D. MacDonald 1973. Transformation of African and Indian family traditions in the Southern Caribbean. *Compar. Stud. Soc. Hist.* **15**, 171–98.

Mahabir, W. 1978. *In and out of politics*. Port of Spain.

Malik, Y. K. 1971. *East Indians in Trinidad: a study in minority politics*. London, New York and Toronto: Oxford University Press.

Mamak, A. 1978. *Colour, culture and conflict: a study of pluralism in Fiji*. Oxford, Toronto and Paris: Pergamon Press.

Matthews, B. 1953. *Crisis in the West Indian family*. University College of the West Indies: Extra-mural Department.

Mayer, A. C. 1961. *Peasants in the Pacific: a study of Fiji Indian rural society*. London: Routledge & Kegan Paul.

Mayer, A. C. 1963. *Indians in Fiji*. London, Melbourne and Bombay: Oxford University Press.

Mischel, W. 1961. Delay of gratification, need for achievement and acquiescence in another culture. *J. Abn. Social Psychol.* **62**, 543–52.

Mittelholzer, E. 1964. *A morning at the office*. London: Penguin.

Morton, S. E. 1916. *John Morton of Trinidad*. Toronto: Westminster.

Mulchansingh, V. 1971. The oil industry in the economy of Trinidad. *Carib. Stud.* **11**, 73–100.

Naipaul, Shiva 1970. *Fireflies*. London: Penguin

Naipual, Shiva 1973. *The chip–chip gatherers*. London: André Deutsch.

Naipaul, Supersad 1976. *The adventures of Gurudeva and other stories*. London: André Deutsch.

Naipaul, V. S. 1957. *The mystic masseur*. London: André Deutsch.

Naipaul, V. S. 1958. *The suffrage of Elvira*. André Deutsch.

Naipaul, V. S. 1960. *Miguel Street*. London: André Deutsch.

Naipaul, V. S. 1961. *A house for Mr Biswas*. London: André Deutsch.

Naipaul, V. S. 1962. *The middle passage*. London: André Deutsch.

Naipaul, V. S. 1964. *An area of darkness*. London: André Deutsch.

Naipaul, V. S. 1967a. *The mimic men*. London: André Deutsch.

Naipaul, V. S. 1967b. *A flag on the island*. London: André Deutsch.

Naipaul, V. S. 1972. *The overcrowded barracoon*. London: André Deutsch.

Naipaul, V. S. 1975. *Guerillas*. London: André Deutsch.

Naipaul, V. S. 1980. *The return of Eva Peron with the killings in Trinidad*. London: André Deutsch.

Naipaul, V. S. 1982. Introduction. In *East Indians in the Caribbean*, B. Brereton and W. Dookeran (eds). New York, London and Nendeln: Kraus International.

Naipaul, V. S. 1983. *Among the believers: an Islamic journey*. London: Penguin.

Nevadomsky, J. 1980a. Abandoning the retentionist hypothesis: family changes among East Indians in rural Trinidad. *Int. J. Sociol. Fam.* **10**, 181–98.

Nevadomsky, J. 1980b. Changes over time and space in the East Indian family in rural Trinidad. *J. Compar. Fam. Stud.* **11**, 433–56.

Nevadomsky, J. 1981a. Wedding rituals and changing women's rights among the East Indians in rural Trinidad. *Int. J. Wom. Stud.* **4**, 484–96.

Nevadomsky, J. 1981b. Cultural and structural dimensions of occupational prestige in an East Indian Community in Trinidad. *J. Anthropol. Res.* **37**, 343–59.

Nevadomsky, J. No date. Changing patterns of marriage, family and kinship among the East Indians in rural Trinidad. *Anthropos*.

Nevadomsky, J. 1982a. Social change and the East Indians in rural Trinidad. *Social Econ. Stud.* **31**(1), 90–126.

Nevadomsky, J. 1982b. Changing conceptions of family regulation among the East Indians in rural Trinidad. *Anthropol. Q.* **55**, 189–98.

Nevadomsky, J. 1983. Economic organization, social mobility, and changing social status among East Indians in rural Trinidad. *Ethnology* **22**(1), 63–79.

Nevadomsky, J. No date. Social change and the East Indians in rural Trinidad: a critique of methodologies. Benin-City, Nigeria. Unpublished.

Nevill, H. R. 1909. *Gorakhpur: a gazetteer*. Allahabad: Government Press.

Nicholls, D. G. 1971. East Indians and black power in Trinidad. *Race* **12**, 443–59.

Nicholls, D. 1979. *From Dessalines to Duvalier*. Cambridge, London and New York: Cambridge University Press.

Niehoff, A. No date. The survival of Hindu institutions in an alien environment. *Eastern Anthrop*. 12.

Niehoff, A. 1958. East Indians of the West Indies. *Lore* **8**, 106–13; **9**, 2–9.

Niehoff, A. 1967. The function of caste among the Indians of the Oropuche Lagoon, Trinidad. In *Caste in overseas Indian communities*, B. M. Schwartz (ed), 149–63. San Francisco: Chandler.

Niehoff, A. and J. Niehoff 1960. *East Indians in the West Indies*. Publications in Anthropology, no. 6. Milwaukee: Public Museum.

Osuji, R. C. 1982. *Sociology of education: with a case study of social class and educational achievement in Trinidad*. St Augustine: Institute of Social and Economic Research.

Ottley, C. R. 1971. *The story of San Fernando*. Port of Spain.

Oxaal, I. 1968. *Black intellectuals come to power: the rise of Creole nationalism in Trinidad and Tobago*. Cambridge, Mass: Schenkman.

Oxaal, I. 1971. *Race and revolutionary consciousness*. Cambridge, Mass: Schenkman.

Parris, C. D. 1981. Trinidad and Tobago – September to December 1973. *Soc. Econ. Stud.* **30**, 42–62.

Parsons, T. 1952. *The social system*. London: Tavistock.

Proudfoot, M. J. 1950. *Population movements in the Caribbean*. Port of Spain: Caribbean Commission.

Ramdin, R. 1982. *From chattel slave to wage earner: a history of trade unionism in Trinidad and Tobago*. London: Martin, Brian & O'Keefe.

Ramesar, M. D. 1976a. Patterns of regional settlement and economic activity by immigrant groups in Trinidad: 1851–1900. *Soc. Econ. Stud.* **25**, 187–215.

Ramesar, M. 1976b. The impact of Indian immigrants on colonial Trinidad society. *Carib. Q.* **22**(1), 5–18.

Ramesar, M. D. 1978. Factors in the settlement of the Indians in Trinidad: 1921–46. In *Actes du 52e congrès international des Américanistes* no. 1, 79–96. Paris.

Rampersad, F. No date. *Growth and structural change in the economy of Trinidad and Tobago, 1951–61*. Jamaica: Institute of Social and Economic Research, University of the West Indies.

Reynolds, J. 1971. Family planning dropouts in Trinidad and Tobago. *Soc. Econ. Stud.* **22**, 176–87.

Richardson, B. C. 1975. Livelihood in rural Trinidad in 1900. *Annals of the Association of American Geographers* **65**, 240–51.

Rienzi, A. C. 1932. *The Beacon*. Quoted in Brinsley Samaroo, 1974. Politics and Afro-Indian relations in Trinidad. In *Calcutta to Caroni*, J. La Guerre (ed.), 84–97. London: Longman.

Roberts, G. W. and L. Braithwaite 1962. Mating among East Indian and non-Indian women in Trinidad. *Soc. Econ. Stud.* **11**, 203–35.

Roberts, G. W. and L. Braithwaite. 1967a. Fertility differentials in Trinidad. *Centr. Stat. Off. Res. Paps* **4**, 95–101.

Roberts, G. W. and L. Braithwaite 1967b. Fertility differentials by family type in Trinidad. *Centr. Stat. Off. Res. Paps* **4**, 102–19.

Roberts, G. W. and L. Braithwaite 1967c. A gross mating table for a West Indian population. *Cent. Stat. Off. Res. Paps* **4**, 128–47; reprinted in 1961 from *Pop. Stud.* **14**(3).

Roberts, G. W. and J. Byrne 1966. Summary statistics on indenture and associated migration affecting the West Indies, 1834–1918. *Pop. Stud.* **20**, 125–34.

Robins, B. 1967. *Trinidad and Tobago: isles of the immortelles*. London: Faber.

Robinson, A. N. R. 1971. *Patterns of political and economic transformation in Trinidad and Tobago*. Cambridge, Mass: Massachusetts Institute of Technology Press.

Robinson, V. 1984. Asians in Britain: a study in encapsulation and marginality. In *Geography and ethnic pluralism*, C. Clarke, D. Ley and C. Peach (eds), 231–57. London: George Allen & Unwin.

Robinson, W. S. 1950. Ecological correlation and the behaviour of individuals. *Am. Sociol. Rev*. **15**, 351–57.

Rodman, H. 1971. *Lower-class families: the culture of poverty in Negro Trinidad*. New York, London and Toronto: Oxford University Press.

Rodney, W. 1981. *A history of the Guyanese working people, 1881–1905*. Baltimore and London: Johns Hopkins University Press.

Rubin, V. 1959. Approaches to the study of national characteristics in a multicultural society. *Int. J. Social Psychiat*. **5**, 20–6.

Rubin, V. (ed.) 1960. *Social and cultural pluralism in the Caribbean. Ann. N.Y. Acad. Sci*. **83**. New York: New York Academy of Sciences.

Rubin, V. 1962. Culture, politics and race relations. *Soc. Econ. Stud*. **11**, 433–55.

Rubin, V. and M. Zavalloni 1969. *We wish to be looked upon: a study of the aspirations of youth in a developing society*. New York: Teachers College, Columbia University.

Ryan, S. 1966. The struggle for Afro-Indian solidarity in Trinidad and Tobago. *Trinidad Tobago Index* **4**, 3–28.

Ryan, S. D. 1972. *Race and nationalism in Trinidad and Tobago: a study of decolonization in a multi-racial society*. Toronto: University of Toronto Press.

Ryan, S. D. 1979. Trinidad and Tobago: the general elections of 1976. *Carib. Stud*. **19**(1 & 2), 5–32.

Ryan, S. 1981. The church that Williams built: electoral possibilities in Trinidad and Tobago. *Carib. Rev*. **10**, (2), 12–13, 45–6.

Ryan, S. No date. The disunited labour front. Supplement to *Carib. Stud*.

Ryan, S., E. Greene and J. Harewood 1979. *The confused electorate: a study of political attitudes and opinions in Trinidad and Tobago*. St Augustine, Trinidad: Institute of Social and Economic Studies.

Samaroo, B. 1972. The Trinidad working men's association and the origins of popular protest in a crown colony. *Soc. Econ. Stud*. **21**, 205–22.

Samaroo, B. 1974. Politics and Afro-Indian relations in Trinidad. In *Calcutta to Caroni*, J. La Guerre (ed.), 84–97. London: Longman.

Samroo, B. 1975. The Presbyterian Canadian Mission as an agent of integration in Trinidad during the nineteenth and twentieth centuries. *Carib. Stud*. **14**(4), 41–55.

Samaroo, B. 1976. The making of the 1946 Trinidad constitution. *Carib. Stud*. **15**(4), 5–27.

Samaroo, B. 1982. Missionary methods and local responses: the Canadian Presbyterians and the East Indians in the Caribbean. In *East Indians in the Caribbean*, B. Brereton and W. Dookeran (eds), 93–115. New York, London and Nendeln: Kraus International.

Sandoval, J. M. 1983. State capitalism in a petroleum-based economy: the case of Trinidad and Tobago. In *Crisis in the Caribbean*, F. Ambursley and R. Cohen (eds), 247–68. Kingston, Port of Spain and London: Heinemann.

Schwartz, B. 1964a. Patterns of East Indian family organization in Trinidad. *Carib. Stud*. **5**, 23–36.

Schwartz, B. M. 1964b. Ritual aspects of caste in Trinidad. *Anthropol. Q*. **37**, 1–15.

Schwartz, B. M. 1964c. Caste and endogamy in Trinidad. *Southwest. J. Anthropology* **20**, 58–66.

Schwartz, B. M. 1965. Extra-legal activities of the village pandit in Trinidad. *Anthropol. Q*. **38**(2), 62–71.

Schwartz, B. M. 1967a. The failure of caste in Trinidad. In *Caste in overseas Indian Communities*, B. M. Schwartz (ed.), 117–47. San Francisco: Chandler.

Schwartz, B. M. 1967b. Differential social and religious adaptation. *Soc. Econ. Stud.* **16**, 237–48.
Seers, D. 1969. A step towards a political economy of development: Trinidad and Tobago. *Soc. Econ. Stud.* **18**, 217–53.
Selvon, S. 1958. *Turn again tiger*. London: MacGibbon & Kee.
Selvon, S. 1963. *I hear thunder*. London: MacGibbon & Kee.
Selvon, S. 1970. *The plains of Caroni*. London: MacGibbon & Kee.
Selvon, S. 1971. *A brighter sun*. Trinidad and Jamaica: Longman.
Sewell, W. G. 1861. *The ordeal of free labour in the British West Indies*. London: Sampson Low.
Shevky, E. and W. Bell, 1955. *Social area analysis*. Stanford, California: Stanford University Press.
Simmons, A. S. 1982. *Modern Mauritius*. Bloomington: Indiana University Press.
Simpson, G. E. 1964. *The Shango cult in Trinidad*. Rio Piedras, Puerto Rico: Institute of Caribbean Studies.
Simpson, J. M. 1973. *Internal migration in Trinidad and Tobago*. University of the West Indies: Institute of Social and Economic Research.
Singaravélou. 1975. *Les Indiens de la Guadeloupe*. Bordeaux: Imprimerie Deniaud.
Singh, H. P. 1965. *The West Indian enigma: a review of C. L. R. James' 'West Indians of East Indian descent or a study of coolietude'*. Port of Spain.
Singh, K. 1974. East Indians and the larger society. In *Calcutta to Caroni*, J. La Guerre (ed.), 39–68. London: Longman.
Singh, K. 1982. Adrian Cola Rienzi and the labour movement in Trinidad (1925–44). *J. Carib. Hist.* **16**, 10–35.
Smith, M. G. 1962. *West Indian family structure*. Seattle: University of Washington Press.
Smith, M. G. 1965a. *The plural society in the British West Indies*. Berkeley, Los Angeles and London: University of California Press.
Smith, M. G. 1965b. *Stratification in Grenada*. Berkeley, Los Angeles and London: University of California Press.
Smith, M. G. 1966. A structural approach to comparative politics. In *Varieties of political theory*, D. Easton (ed.), 113–28. Englewood Cliffs, N.J.: Prentice-Hall.
Smith, M. G. 1969. Some developments in the analytic framework of pluralism. In *Pluralism in Africa*, L. Kuper and M. G. Smith (eds), 415–58. Berkeley, Los Angeles and London: University of California Press.
Smith, M. G. 1974. *Corporations and Society*. London: Duckworth.
Smith, M. G. 1984a. *Culture, class and race in the Commonwealth Caribbean*. Mona, Jamaica: Department of Extra-Mural Studies, University of the West Indies.
Smith, M. G. 1984b. The nature and variety of plural units. In *The prospects for plural societies*, D. Maybury-Lewis (ed.), 146–86. Washington D.C.: American Ethnological Society.
Smith, R. J. 1963. *Muslim East Indians in Trinidad: retention of ethnic identity under acculturative conditions*. University of Pennsylvania, unpublished PhD thesis.
Smith, R. T. 1961. Review of *Social and cultural pluralism in the Caribbean*, V. Rubin (ed.). Ann. N.Y. Acad. Sci. **83**. New York: New York Academy of Sciences. In *Am. Anthrop*. **63**, 155.
Smith, R. T. and C. Jayawardena, 1958. Hindu marriage customs in British Guiana. *Soc. Econ. Stud.* **7**, 178–94.
Smith, R. T. and C. Jayawardena 1959. Marriage and family amongst East Indians in British Guiana. *Soc. Econ. Stud.* **8**, 321–76.
Smith, R. T. and C. Jayawardena 1967. Caste and social status among the Indians of Guyana. In *Caste in overseas Indian communities*, B. M. Schwartz (ed.), 43–92. San Francisco: Chandler.
Solomon, P. 1981. *Solomon: an autobiography*. Port of Spain.

Speckmann, J. D. 1965. *Marriage and kinship among the Indians in Surinam*. Assen: Van Gorcum.

Stone, C. 1972. *Stratification and political change in Trinidad and Jamaica*. Beverly Hills, California: Sage.

Sutton, P. K. 1982. Dr. Eric Williams and politics in Trinidad. *Carib. Soc.* no. 1. Institute of Commonwealth Studies, London University.

Sutton, P. K. 1983. Black power in Trinidad and Tobago: the crisis of 1970. *J. Commonwlth Compar. Polit.* **21**, 115–32.

Sutton, P. 1984. Trinidad and Tobago: oil capitalism and the presidential power of Eric Williams. In *Dependency under challenge*, A. Payne and P. Sutton (eds), 43–76. Manchester: Manchester University Press.

Thomas, J. J. 1969. *Froudacity: West Indian fables explained*. London and Port of Spain: New Beacon.

Thomas, R. D. 1972. *The adjustment of displaced workers in a labour-surplus economy*. Jamaica: Institute of Social and Economic Research, University of the West Indies.

Thomas-Hope, E. 1980. The pattern of Caribbean religions. In *Afro-Caribbean religions*, B. Yates (ed.), 4–15. London: Ward Lock.

Tikasingh, G. 1982. Toward a formulation of the Indian view of history: the representation of Indian opinion in Trinidad, 1900–21. In *East Indians in the Caribbean*, B. Brereton and W. Dookeran (eds) 11–32. New York, London and Nendeln: Kraus International.

Tinker, H. 1974. *A new system of slavery: the export of Indian labour overseas 1830–1920*. London, New York, Bombay: Oxford University Press.

Tinker, H. 1982. British policy towards a separate Indian identity in the Caribbean, 1920–50. In *East Indians in the Caribbean*, B. Brereton and W. Dookeran (eds), 33–47. New York, London and Nendeln: Kraus International.

Trinidad and Tobago fertility survey 1977: a summary of findings. *World Fertil. Surv.* **33**, 1–14.

Tyson, J. D. 1939. *Report on the condition of Indians in Jamaica, British Guiana and Trinidad*. Simla: Government of India Press.

UNESCO 1977. *Race and class in post-colonial society: a study of ethnic group relations in the English-speaking Caribbean, Bolivia, Chile and Mexico*. Paris: UNESCO.

de Verteuil, L. A. A. 1884. *Trinidad*, 2nd edn. London: Cassell.

Vertovec, S. A. 1984. The East Indians of Trinidad: their social, cultural and religious context. Faculty of Anthropology and Geography, Oxford University. Unpublished.

Wagley, C. 1960. Discussion of M. G. Smith's paper, Social and cultural pluralism. In *Social and cultural pluralism in the Caribbean*, V. Rubin (ed.). Ann. N.Y. Acad. Sci. **83**, 777–80. New York: New York Academy of Sciences.

Weller, J. A. 1968. *The East Indian indenture in Trinidad*. Rio Piedras, Puerto Rico: Institute of Caribbean Studies.

Williams, E. 1962. *History of the people of Trinidad and Tobago*. Port of Spain: Peoples National Movement.

Williams, E. 1969. *Inward hunger: the education of a Prime Minister*. London: André Deutsch.

Williams, E. 1981. *Forged from the love of liberty: selected speeches compiled by Paul K. Sutton*. Trinidad: Longman.

Wood, D. 1968. *Trinidad in transition*. London and New York: Oxford University Press.

Woods, R. I. 1976. Aspects of the scale problem in the calculation of segregation indices: London and Birmingham 1961 and 1971. *Tijdschr. Econ. Soc. Geograf.* **67**(3), 169–74.

Yawney, C. D. 1969. Drinking patterns and alcoholism in Trinidad. In *McGill studies in Caribbean anthropology*, F. Henry (ed.), 34–48. Occasional Paper Series, no. 5. Montreal: McGill University, Centre for Developing Area Studies.

Zavaloni, M. 1968. *Adolescents' values in a changing society: a study of Trinidad youth*. The Hague and Paris: Mouton.

Index

Italic type refers to text figures.

acculturation 149
Agor Panth 98
All Saints Night 107
All Trinidad Sugar Estates and Factory Workers' Union 22
Alliance 147, 149
Anglicisation 37
Anjuman Sumat Al Jamaat (ASJA) 99, 101, 102, 103
  ASJA College 103
Arya Samaj 98, 100
  wedding 123

Bhagavad-gita 89, 103
Bhagwad Yagya 101
blacks 18, 35, 59, Table 3.1, *2.5*
  see also Negro
black power 146, 151
Bombay Indians 36, 54
Brahmin 16
  priesthood and control of religion 104–5
  purity and pollution 104
  see also caste
Braithwaite, L. E. S. 23, 147
Bramha 100
brown 18, 44
  see also mixed and coloured
Buckra Eid 108
Butler, T. U. (Buzz) 22, 43
Butler Party 22

Calcutta 8, 10
Canadian Mission Indian Schools 17, 37, 61, 99
  Grant Memorial School 37, *3.1*
Capildeo, Dr Rudrinath 136, 139
Carib 58, Table 4.3
Carnival 105–7
  attitudes among samples 106–7
  costume wearers 106
  jouvert 106
  ole mas 106
  participation among samples 106
  steel bandsmen 106
caste 10, 89–93
  among Christian East Indians 96–7, 127
  and class 92–7, Table 5.6, Table 5.7
  erosion 14
  religion and priesthood 103–5
  varna and endogamy 124–7, Table 7.3
  hypergamy and hypogamy 126
  in Débé 96–7
  jat 89–92
  varna 89–92
census data 152
Chambers, George 147
Chinese 8, 36, 44, 58, 59, 63, 76, Table 3.1, Table 4.1, Table 4.3, Table 4.4, Table 4.5, *2.5*, *4.1*, *4.6*, *4.7*, *4.13*
Christian East Indians 4, 6, 18, 44, *2.4*, *2.5*, *4.2*, *4.5*, *4.6*, *4.8*, *4.13*
  attitude to carnival 106–7
  caste 96–7, Table 5.6
    varna and endogamy 127
  courtship and marriage 121
  education 82, Table 5.3
  employment 84
  film watching 80
  friendship and clubbing 133–6
  geographical mobility 80
  household headship, marital condition and parental status 113–15, Table C.1, Table C.2
  housing 85–6
  income 84–5
  intermarriage 129–33, Table 8.1
  involvement in Moslem events 107–9
  occupation 81, Table 5.1
  occupational
    aspirations for children 83
    mobility 84–5
  ownership of consumer durables 86
  participation in
    All Saints Night 107
    carnival 106
    Diwali 107
    other Hindu festivals 107
    Siparu Mai 111
  placement of children 118–19, Table 7.2
  political affiliation 135–42, Table 8.2, Table 8.3
  Presbyterian marriage 123–4
  race and access to jobs 86–8, Table 5.4, Table 5.5
  racial attitudes 87–9
  relationship to household head 116, Table C.3
  religious affiliation 98–9
  type of domestic unit 117–18, Table 7.1
  unemployment 84
  union status 119–21
  see also Presbyterian
Christianity *4.2*, *4.4*, *4.5*, *4.8*, *4.11*
  conversion to 101–2

see also syncretism, spirit world, festivals
Christian sects and cults 69, *4.4*, *4.11*
see also syncretism
Church of the Open Bible 99, 103
class 1, 27, *4.9*
and voting 140
colour class 58, 77, *2.5*
consensus 27
occupations in 1946 Table 3.2
see also stratification
club membership 134–6
colour and occupation 36, Table 5.2, *4.13*
coloured *2.5*
free coloureds 35
see also mixed and brown
corporate
categories 26
groups 26
Creole–East Indian
convergence politically 140
hostility 36
pluralism 57–8
polarisation 63
Creoles 3, 4, 6, 134, Table 4.4
attitude to carnival 106–7
British Creole 44
colour stratification 80–1
courtship and marriage 121
education 82, Table 5.3
employment 84
film watching 80
French Creole 44
friendship and clubbing 133–6
geographical mobility 80
household headship, marital condition and parental status 112–13, Table C.1, Table C.2
housing 85–6
income 84–5
intermarriage 129–33, Table 8.1
involvement in Moslem events 108–9
non-white Creoles 48
occupation 82, Table 5.1, Table 5.2
occupational
aspirations for children 84
mobility 84–5
ownership of consumer durables 86
participation in
All Saints Night 107
Carnival 106
Diwali 107
other Hindu festivals 107
Siparu Mai 111
placement of children 118–20, Table 7.2
political affiliation 135–42, Table 8.2, Table 8.3
race and access to jobs 86–8, Table 5.4, Table 5.5
racial attitudes 88–9
relationship to household head 115–16, Table C.3
religious affiliation 98
type of domestic unit 117–18, Table 7.1
unemployment 84
union status 119–21
Creolise 144
Crowley, D. 24, 109
cultural category 1

Dass, Tulsi 16, 100
Débé 6, 12, 77, *2.1*
attitude to carnival 106–7
caste 96–7, Table 5.6
varna and endogamy 127
courtship and marriage 121
education 82, Table 5.3
employment 84
film watching 80
friendship and clubbing 133–6
geographical mobility 80
household headship, marital condition and parental status 113–15, Table C.1, Table C.2
housing 85–6
income 85
intermarriage 129–33, Table 8.1
involvement in Moslem events 107–9
marriage 120–3
occupation 81, Table 5.1
occupational
aspirations for children 84
mobility 84–5
ownership of consumer durables 86
participation in
All Saints Night 107
carnival 106
Diwali 107
other Hindu festivals 107
Siparu Mai 111
placement of children 118–20, Table 7.2
political affiliation 135–41, Table 8.2, Table 8.3
race and access to jobs 86–8, Table 5.4, Table 5.5
racial attitudes 88–9
relationship to household head 116, Table C.3
religious affiliation 99
type of domestic unit 117–18, Table 7.1
unemployment 84
union status 119–21
Democratic Action Congress 147
Democratic Labour Party (DLP) 23, 135–41, 147
de Verteuil, L. 32
Diwali 107
Douglas 4, 6, 15, 58, 88, 145, Table 4.2
attitude to carnival 106–7
courtship and marriage 121

education 82, Table 5.3
employment 84
film watching 80
friendship and clubbing 133–6
geographical mobility 80
household headship, marital condition and parental status 113, Table C.1, Table C.2
housing 85
income 85
intermarriage 129–33, Table 8.1
involvement in Moslem events 108–9
occupation 81, Table 5.1
occupational
aspirations for children 84
mobility 84–5
ownership of consumer durables 86
participation in
All Saints Night 107
carnival 106
Diwali 107
other Hindu festivals 107
Siparu Mai 111
placement of children 118–21, Table 7.2
political affiliation 135–41, Table 8.2, Table 8.3
race and access to jobs 86–8, Table 5.4, Table 5.5
racial attitudes 88–9
relationship to household head 115, Table C.3
religious affiliation 98
type of domestic unit 117–18, Table 7.1
unemployment 84
union status 119–21

East Indians Table 3.1, Table 4.1, Table 4.3, Table 4.4, Table 4.5, *2.3*, *2.5*, *4.2*, *4.5*, *4.6*, *4.8*, *4.13*
attachment to ancestral home 79
clothes 78
family 14–15, *4.10*
food 78
interest in cinema 80
National Association 21
National Congress 21, 22
retailing 45, Plate 1
support for PNM 44, 140
see also Hindus, Moslems, Christian East Indians (Presbyterian)
educational standard 69, Table 4.5, *4.12*, *4.13*
Eid ul Fitr 108
elite 44–7
dentistry 45
law 45
medicine 45
professions 45–6
ethnic group 1

festivals 105–9
Fiji 5, 150
Fitzpatrick, G. 21
friendship 133–4
Froude, J. A. 16, 21
Furnivall, J. S. 25

Genovese, E. 28
Guyana 5, 150–1

Hanuman 16, 100, 101
Harris Promenade 38, *3.1*
Hindi 14, 16–17, 21, 29, 83, 102
Hinduism 99
Brahmin control of priesthood and Sanathanism 104–5
cremation 110, Plate 2
institutional revival 102–3
marriage 120–3, Plate 4, Plate 5, Plate 6, Plate 7
purity and pollution 104
reincarnation 104
vegetarianism 104
see also caste, festivals, puja, spirit world, syncretism
Hindus 4, 6, 18, 29, 51, 58, 61, 63, Table 4.1, Table 4.2, Table 4.3, Table 4.4, Table 4.5, *2.4*, *2.5*, *4.2*, *4.6*, *4.8*, *4.13*
attitude to carnival 106–7
caste
and class 92–7, Table 5.6, Table 5.7
varna and endogamy 124–7, Table 7.3
courtship and marriage 121
education 82, Table 5.3
employment 84
film watching 80
friendship and clubbing 133–6
geographical mobility 80
Hindu–Presbyterian (Christian) link 65, 145
household headship, marital condition and parental status 113–15, Table C.1, Table C.2
housing 85–6
income 85
intermarriage 129–33, Table 8.1
involvement in Moslem events 108–9
jat 89–92
marriage 120–3, Plate 4, Plate 5, Plate 6, Plate 7
occupation 82, Table 5.1
occupational
aspirations for children 84
mobility 84–5
ownership of consumer durables 86
participation in
All Saints Night 107
carnival 106
Diwali 107
other Hindu festivals 107
Siparu Mai 111

placement of children 118–20, Table 7.2
political affiliation 135–42, Table 8.2, Table 8.3
race and access to jobs 86–8, Table 5.4, Table 5.5
racial attitudes 88–9
relationship to household head 116, Table C.3
religious affiliation 98–9
type of domestic unit 117–18, Table 7.1
unemployment 84
union status 119–21
varna 89–93
Holi (Phagwa) 108
Hosay 108–9
riot of 1884 109
Tadjahs 108

illness 109
jharay 109
incorporation
differential 26, 29
equivalent (or segmental) 26
uniform (or universalistic) 26
index of dissimilarity 51, 64, Table 4.1, Table 4.4, Table 9.1, *2.3*, *2.4*
see also segregation
Indian
indentured
immigration 8, 10
labour 12–13
population 9, *4.5*, *4.6*
provenance 9–10
Islam 99
five pillars (kalima, namaaz, roza, haj and zakat) 101
institutional revival 102, 103
see also syncretism and festivals
intermarriage 129–33, Table 8.1

janeo 104
jhandi 100, 149
Joseph, Roy 43, 148

Kabir Panth 98, 100
Kali Mai 16
Kalkatiyas 14
Kartik Nahan 108
Klass, M. 23, 24, 92
Kuper, L. 28

labour force 67, Table 3.2, *4.9*
linkage analysis 56, 69–75, *4.6*, *4.13*
Lowenthal, D. x, 24, 142, 150

Madeira 8
see also Portuguese
Madras 8, 10, 17
Madrassi fire walking 16
Mahabir, Dr W. 43
Maraj, B. 22, 47, 135, 142
Marxist 1, 28
Mauritius 5, 150
Mittelholzer, E. 87
mixed 59, 61, Table 4.3, Table 4.4, Table 4.5, *2.3*, *4.7*, *4.13*
see also brown and coloured
Mohammed, S. 44
Montano, G. 44
Morton, Rev. John 17, 37
Moslems 4, 6, 10, 18, 49, 58, 61, Table 4.1, Table 4.2, Table 4.3, Table 4.4, Table 4.5, *2.4*, *2.5*, *4.2*, *4.6*, *4.8*, *4.13*
attitude to carnival 106–7
courtship and marriage 121
education 82, Table 5.3
employment 84
film watching 80
friendship and clubbing 133–6
geographical mobility 80
household headship, marital condition and parental status 113–15, Table C.1, Table C.2
housing 85–6
income 85
intermarriage 129–33, Table 8.1
involvement in Moslem events 108–9
occupation 82, Table 5.1
aspirations for children 84
mobility 84–5
ownership of consumer durables 86
participation in
All Saints Night 107
carnival 16
Diwali 107
other Hindu festivals 107
Siparu Mai 111
placement of children 118–20, Table 7.2
political affiliation 135–42, Table 8.2, Table 8.3
race and access to jobs 86–8, Table 5.4, Table 5.5
racial attitudes 88–9
relationship to household head 116, Table C.3
religious affiliation 98–9
type of domestic unit 117–18, Table 7.1
unemployment 84
union status 119–21
wedding 123
mosque 37, 63
Friday worship 101

Naipaul, V. S. 3, 20, 23, 37, 47, 79, 85, 93, 101, 144
Naparima
College 42, *3.1*
Girls' School 42, *3.1*

Naparimas 18, 32–4, 41
Negro Table 4.3, Table 4.4, Table 4.5, *2.3*, *4.7*, *4.13*
see also black
Niehoff, A. and J. 23
non-residential land use 38

obeah man 109
oil 41, 147
Oilfield Workers' Trade Union 22
ojha man 109
Organisation for National Reconstruction 147
Oropuche Lagoon 6, 11, 23, *2.1*

panchayat 14
Penal 12
People's Democratic Party (PDP) 22, 136
origin in Sanathan Dharma Maha Sabha 145
People's National Party (PNM) 22, 136–41, 147
pluralism
cultural 25–6, 148, 150
model of *2.5*
origins of 29
social 26, 148, 150
structural 26, 148, 150
Pointe-à-Pierre 41
politics 43–4, 135–41, Table 8.2, Table 8.3
see also Alliance, Butler Party, Democratic Action Congress, Democratic Labour Party, Organization for National Reconstruction, Peoples Democratic Party, Peoples National Movement, Tapia, United Labour Front, Workers and Farmers Party
population and housing 39, 51
distribution of population in San Fernando *3.2*
housing types in San Fernando *4.3*, *4.6*
Port of Spain 8, 151, *2.1*
Portuguese 58, Table 4.3
see also Madeira
Presbyterian 17, 37, 49, 58, 61, Table 4.1, Table 4.2, Table 4.3, Table 4.4, Table 4.5, *4.2*, *4.6*, *4.8*, *4.13*
acculturation 44
mixed marriage 124
retention of Hindu values 102
Susamachar Church and conversion 102
Susamachar Church in San Fernando 37
wedding 123–4
Presentation College 42, *3.1*
puja 100–1, 102

race 1–2
racial groups and their characteristics 74–6
distribution
in Trinidad 2.3
in San Fernando *4.1*, *4.7*
1946 population in San Fernando Table 3.1
race and religion Table 4.2, Table 4.3, Table 4.5
see also Creoles, Douglas, East Indians
racial stereotyping 46
racism 1, 21, 28, 88–9, 150
Rama 100, 101
Ramadan 108
Ramandi Panthi 100
see also Sanathan Dharma
Ramayana 16, 100
Rameyn Yagya 101
Ramlila 107
Ram Naumi 107
Ramsarran, V. C. 43
religion 16–18
Christian religions in San Fernando *4.4*, *4.11*
cultural indicators in San Fernando *4.5*
distribution of groups in Trinidad *2.4*
East Indian religious categories in San Fernando *4.2*, *4.8*
in San Fernando 37–8, 51–4, 69
race and religion Table 4.3
see also Christian East Indians, Hindus and Moslems
Rienzi, A. C. 21–2, 43
Robinson, A. N. R. 148
Roodal, T. 21–2, 43

St Joseph's Convent 42, *3.1*
sample survey 154
size of samples 154
Sanathan Dharma 16, 22, 37, 100, 102, 145
see also Ramanandi Panthi
San Fernando place names *3.1*
segment 77
segmentation 78–80
segregation 67, 143–4, 146–7, *2.3*, *2.4*
scale factors in 144, Table 4.1, Table 4.4, Table 9.1
see also index of dissimilarity
Selfless Service Divine Mission 102
Seunerine 98, 100
sex ratio in San Fernando 35, Table 3.1
Shia 99, 108
Siparu Mai 110–11, Plate 3
Sita 100, 101
Siva 100
lingam 101
Siw Ratvi 107
Skinner Park 39, *3.1*
Smith, M. G. x, 25, 26–8, 29, 69–73, 115, 148
social
section 27
segment 27
spirit world 109
jumbies 109
lagahoos 109
maljeu 109

soucouyants 109
stratification 78, 80–6
 occupations in 1946 Table 3.2
 see also class
Sunni 98, 108
Surinam 5, 150
syncretism 109–11
Syrians 36, 58, Table 4.3

Takveeyatul Islamic Association (TIA) 101
Tapia 147
temple 16, 37, 63
Thomas, J. J. 42
town plan 37–8, *3.1*
Trinidad Defence Force 149
Trinidad Leasehold 41
Trinidad Moslem League (TML) 98, 101

union status 67, 75
United Labour Front (ULF) 147, 149
Urdu 16, 21, 83, 102
Usine Ste Madeleine 39, 41, 44

Vishnu 16, 100

Weber, M. 1
Weller, J. 12
White 18, 35, 48, 58, 61, Table 3.1, Table 4.1, Table 4.2, Table 4.3, Table 4.4, Table 5.2, *2.3*, *2.5*, *4.1*, *4.6*, *4.7*
Williams, Dr Eric 3, 22, 24, 136, 147, 149
Workers and Farmers Party 145

For Product Safety Concerns and Information please contact our EU representative GPSR@taylorandfrancis.com Taylor & Francis Verlag GmbH, Kaufingerstraße 24, 80331 München, Germany

Printed and bound by CPI Group (UK) Ltd, Croydon, CR0 4YY

08/06/2025

01897011-0007